Biology:
A Search for Order in Complexity
Second Edition

TEACHER'S MANUAL

68911 R6/08

A publication of
Christian Liberty Press
502 West Euclid Avenue
Arlington Heights, Illinois 60004
www.christianlibertypress.com

Written by: Edward J. Shewan
             Diane C. Olson
             Lars R. Johnson

Editing by: Edward J. Shewan
Copyediting by: Diane C. Olson

ISBN 978-1-930367-93-7
       1-930367-93-7

Printed in the United States of America

CONTENTS

Preface

TWO DIFFERENT APPROACHES

Two different approaches can be used in preparing a science textbook. By **one approach**, the data of science would simply be presented as it is, with accurate descriptions of the pertinent systems and processes and their interrelationships in the present world. This approach is true **science**, which is "knowledge"—*the organized body of observation of, and experiment with, present processes*. In biology, this method would entail careful descriptions of the plant, animal, and human worlds and the complex processes that constitute the phenomenon of life.

But most of us are not content with merely knowing about things as they now exist. We are also intensely interested in how they came to be as they are. Thus, authors of most textbooks of science follow the second approach, seeking not only to describe the phenomena of a particular science as they now function but also, if possible, to explain how they got that way.

In biology, the **second approach** necessarily entails a philosophic viewpoint regarding **origins**. We want to know not only the characteristics of the various living organisms but also when these organisms came into existence. Consequently, practically all biology textbooks include a discussion of the supposed origin of life of the various forms of plants and animals.

But a word of caution is necessary here. The discussion of origins, strictly speaking, is *not* science. This is because origins are not subject to experimental verification. No scientific observers were present when life began or when different kinds of organisms first came into existence, and these events are not taking place in the present world; therefore, the solution to the question of origins is simply impossible by scientific means. *The student should always be careful* (though some textbook writers are not) *to distinguish between the actual facts of biology and some viewpoint about origins with which particular biologists may try to explain those facts.*

There are essentially only two philosophic viewpoints of origins among modern biologists—the doctrine of **evolution** and the doctrine of **special creation**. Proponents of the former postulate the gradual appearance of the various forms of life and of life itself by natural processes over vast ages of time. Exponents of the latter assume the essentially instantaneous origin of life and of the major kinds of living organisms by special creative acts utilized directly by the Creator Himself.

Both evolutionists and creationists agree to the actual facts of biology, but the ***interpretation*** of those facts in relation to the question of origins and intrinsic meanings will depend upon the philosophic preference of the individual biologist.

Scientists work in terms of "models," and each proposed model is evaluated in terms of the effectiveness with which available data may be correlated into the model as a frame of reference. Following this example, the two basic viewpoints of origins may be called the "evolution model" and the "creation model." A choice between these two models may be made in terms of the effectiveness with which each may be used to correlate available data.

At this point it may be well to recognize that some persons have attempted to formulate a **middle position** of some kind between the two models, by which they hope to accept *both* evolution and creation. That is, they hold that perhaps evolution was the *method* of creation. The tenability of this type of compromise can best be considered after the two basic models are first evaluated. In any case, it is certainly true that many evolutionist scientists and many creationist scientists alike reject this idea.

The consistent evolutionist insists that if indeed innate evolutionary processes suffice to explain all the data, as he *believes* they do, then there is no need to invoke extraneous creative processes. The creationist says that, since there is a need to postulate extraneous creative acts and processes to explain the data, as he believes, the evolution model is by that very fact rendered impotent. These two models cannot really be harmonized, except at a very superficial level, since they represent diametrically opposite viewpoints of origins.

THE EVOLUTION MODEL

The present processes of nature are, in the *evolution model*, adequate to explain the origin of the universe and all changes preliminary to the present immense degree of variety and complexity. Despite occasional failures and even retrogressions, the overall effect of these innate principles and processes has been that of the rise of diversity and complexity from primeval simplicity. The processes of the cosmos, therefore, are supposedly processes of origination and integration.

Since, according to this model, all things are interrelated by common descent through slowly operating innate processes, certain **basic predictions** can be made from the model as a test of validity of the model. Consider the following:

1. Countless structural and functional similarities should be observed among the entities of the present world—with, in fact, a more or less continuous array of *inorganic species*, *semi-organic transitional species*, and *organic species*; and with no "gaps" of any consequence between adjacent kinds.

2. The basic processes which have presumably given rise to all things should, when observed in the present world, turn out to be processes which tend to produce new entities in an ever higher state of order and integration.

3. If it is possible to decipher the actual history of the earth, it should be found that the variety and complexity of the world and all its inhabitants tend to increase as time increases.

Support for Predictions Only Apparent

These three predictions may be supported to some degree by the observed data. Many types of similarities are observed between different organisms; for example, similarities in anatomy, in embryonic development, in genetic biochemistry, and in blood serology.

However, the inference of a continuous array of such similarities, with no gaps of any consequence between adjacent kinds, is *not* supported by the data. Although certain conjectures might be offered to explain the existence of the great gaps between all the basic kinds, these ideas are not accessible to experimental test and thus do not afford any genuine scientific explanation for this obvious deficiency in the evolution model.

Secondly, study of various processes does bear out the evolutionary inference that many changes continually take place in the world. In the organic world, for example, new varieties and even new species are easily produced through the mechanisms of hybridization, induced mutation, and selection; and these phenomena may occur either naturally or artificially. No two individuals are alike, even from the same parents, and there is obviously a great deal of variation and change taking place in the world.

Once again, however, this evidence alone is not very compelling, since these processes of change are not innately processes tending toward increase of order as predicted. On the contrary, they seem always to fall into one of two categories:

a) *variation within relatively small limits*, leading merely to new varieties within a basic kind; and

b) *mutations that represent random changes in the DNA of the germ cells*, resulting almost always in decreased order at immature or adult stages and never resulting in the appearance of new physical traits.

Actually, these two phenomena may be used better to support the principles of conservation and decay, rather than origination and integration, as proponents of the evolution model would suggest. These observational fallacies to date have not been overcome by any measurable facts, although evolutionists feel justified in *extrapolating* these small variations into broader evolutionary changes between basic kinds.

Prediction Denied by Basic Facts

The inference that the complexity of life should have increased with the passage of geologic time does *seem* at first to be substantiated by the fossil record; and, indeed, this evidence from paleontology is undoubtedly the strongest of the evidences offered in support of evolution. However, it is seriously weakened by the necessity of **circular reasoning** in its development. That is, the scale of geologic time must necessarily be based on the assumption of evolution in the first place.

The relative dating of the geologic formations is always determined mainly by the "index fossils" that they contain, and their supposed "absolute dating" by radioactive minerals is always subject to correction by these *paleontologic* (pā'lē•ɑn•tə•lɑ'jik) *criteria*. Furthermore, there are many, many locations where fossils from different "ages" are found in the same beds, and even where entire formations containing "old" fossils are superimposed vertically above formations containing "young" fossils.

This argument is still further weakened by the obvious fact that most of the fossiliferous (fɑ'sə•li'fə•rəs) rocks—especially those containing fossils of large plants or animals—must have been deposited and petrified *rapidly*, even **catastrophically**; otherwise, the fossils would not have been preserved at all. Thus, the fossil record does not necessarily reflect slow, uniformitarian evolutionary change over vast ages, but rather it contains a graphic record of violence and death on a worldwide scale.

Though some of the data may possibly be interpreted in an evolutionary framework, this interpretation is not at all compelling or conclusive. The evolution model contains numerous deficiencies and discrepancies. One may adhere to it as an act of *faith*, but it is fallacious and misleading to label it "science."

THE CREATION MODEL

That there was a period of **special creation** in the past, during which the world was brought into existence out of nothing by the power of the Creator, is a basic postulate of the *creation model*. All of the basic physical entities were perfected and all the basic biologic kinds established, each with specific form and function, through intelligent design.

These basic units are now being "conserved" rather than "created." The present processes of nature are therefore not *creative* (nor evolutionary) processes at all, but rather *conservative* processes, which maintain the essential integrity and stability of the universe as created.

This does not mean, of course, that no change or variety is possible. To the contrary, an important postulate of the creation model is that a tremendous complex of *inorganic* and *organic variants* have come into existence from the basic created entities. However, such variation will always be found to be within the limits imposed by the initially created structure of each entity. In the biologic realm, for example, many new varieties (or even species or genera, depending on terminology) may quickly appear in response to environmental constraints, but *never* a new basic "kind."

In addition, according to the biblical version of the creation model, *a universal principle of decay and death* (though not annihilation) throughout the world was established at some time after the creation period. Finally, a great **worldwide cataclysmic flood** at a still later date, which radically changed the face of the earth, as well as the nature and rate of action of most processes on the earth, is incorporated into the creation model.

Specific Predictions Are Confirmed

The above features of the creation model are confirmed by most or all of the *actual observed phenomena* of nature, thus demonstrating the validity of the creation model as being scientifically sound, even though no model of origins can be fully verified by scientific methods.

The two most basic and firmly established scientific principles are the *first* and *second laws of thermodynamics*. These apply without exception to all scientific disciplines and may properly be regarded as confirmed predictions of the creation model. That is, the **first law** (conservation of mass-energy) supports the prediction that nothing is being created or annihilated in the present order of things, since the creation was completed and perfected at some time in the past and is now merely being maintained and conserved. Similarly, the **second law** (increasing entropy) is essentially a confirmation of the universal law of decay and death postulated in accordance with the biblical version of the creation model.

The *permanence of basic "kinds"* is supported without exception by all observed biologic data. Thus, a population of moths may change color because of a change in the smoke content of the atmosphere, but they remain moths. A thousand successive generations of fruit flies may be exposed to radiations and other mutagenic influences, with the production of a wide variety of mutants, but they still are fruit flies.

Great **gaps** between basic kinds are likewise to be expected, since each kind has a created purpose and, therefore, a structure uniquely designed with that purpose in view. On the other hand, many **similarities** would likewise be expected since it is reasonable that, when similar functions are to be performed in similar environments, even different "kinds" would be designed with somewhat similar structures by a common Creator.

As far as the fossil record is concerned, it is well known that essentially the same gaps between basic "kinds" exist in both the fossil record and the present biologic world. There are, of course, many extinct kinds, as well as extinct varieties of present kinds, found in the fossils, but none of these can be considered as actual transitional or *genetic* links between any of the established kinds.

Sedimentary Layers Relate to Cataclysmic Flood

Furthermore, according to the creation model, such fossil deposits should be found in sedimentary beds all over the world. In fact, there seems to be no way of accounting for most of the great fossil beds of the world, especially of vertebrate fossils, except in terms of very rapid burial and *lithification* (lĭ′thə•fə•kā′shən, process by which sediments are consolidated into sedimentary rock), such as might be posited in accordance with the **biblical deluge**, and accompanying volcanic and tectonic activity and inferred subsequent glaciological (glā′shē•ə•lä′jĭ•kəl) phenomena.

The apparent order of "succession" of fossils may be predicted from the creation model to be from the least complex at the bottom to the most complex near the top, though with numerous statistical exceptions to this rule. That is, the *hydrodynamic action* of moving water is a highly efficient sorting agent, and sedimentary contents would be expected to be segregated into aggregations of similar sizes and shapes, normally being deposited in nicely stratified layers.

Hydrodynamically, the simplest (i.e., most nearly spherical) and densest would tend to settle out first and thus be buried deepest. Further, the least complex organisms tend to inhabit the lowest elevations and would, therefore, tend to be deposited at the lowest levels. Finally, the more advanced organisms are the more mobile and would, therefore, be expected to survive floodwaters longer and consequently be trapped and buried at the higher levels if at all.

These should be considered statistical criteria only, of course, and many exceptions would be anticipated in the context of a *universal aqueous cataclysm* lasting an entire year (and, in lesser intensity, for centuries). Both the expected normal sequences and the occasional exceptions are found as predicted in the geologic "column" all over the world.

SUMMARY

Thus, in summary, there are two possible models of origins, the *creation model* and the *evolution model*, though several variants of each have been developed. Both deal with ultimate meanings, and both are incapable of scientific proof. On the other hand, both may be used as frames of reference for development of predictions for comparative evaluation of the present phenomena of nature.

On this basis, the creation model is a framework of interpretation and correlation that is, at least, as satisfactory as the evolution model. However, the two laws of thermodynamics, the apparent stability of the basic "kinds," the existence of great gaps between the kinds, the deteriorative nature of mutations, and the catastrophic nature of the worldwide fossil-bearing formations all may be correlated far more easily with the creation model than with the evolution model. In other words, *it requires more **faith** to believe in evolution rather than special creation.*

Furthermore, the data and principles of physics, chemistry, and the other physical sciences are much more easily understood within the framework of the creation model than in that of the evolution model.

Yet, the majority of modern biologists prefer the evolutionary viewpoint of origins as ***the*** explanation of the factual data of biology. In fact, many have been so confident in this position that *some have insisted that evolution is a **fact** of science*. But this assertion has never been proved and, of course, cannot even be tested.

There also exists today a significant number of biologists and other scientists who are convinced that special creation doctrine or the creation model is a more reasonable and satisfying viewpoint of origins than evolution. Many of these men and women are members of the *Creation Research Society*, an organization of hundreds of scientists (with at least a masters degree in science, and representing most of the disciplines in the physical and biological sciences) dedicated to research and publication in support of creation versus evolution as the most likely explanation of origins.

The preponderance of evolutionists in the present-day scientific and educational establishments, however, has led to an effective monopoly of evolutionary ***opinion*** in modern textbooks. Thus, a great need exists for an introductory biology textbook that can be used effectively both in teaching the actual facts of biologic science and also in presenting the creation concept as the most acceptable underlying explanation of those facts.

John N. Moore, Co-editor of first edition

Foreword

This teacher's manual is designed to assist the teacher in using *BIOLOGY: A Search for Order in Complexity*. Suggestions regarding a correlated laboratory program can be found in the teachers' laboratory manual.

Effectiveness of teaching in biology is influenced by many factors such as academic background, teaching experience, equipment available, and background of students. Since effective teaching depends upon more than subject matter organization, various features of this teacher's manual have been formulated to enhance the interest and enthusiasm with which biological subject matter is presented.

The general design or outline of this teacher's manual parallels the organization of the book as follows:

a. Chapter by chapter suggestions for student motivation and enrichment in regard to the subject matter are presented in interesting and imaginative practical projects.

b. Lists of instructional films and supplemental readings are given for each chapter. The creation viewpoint is presented in some of the films, books, and scientific journals. Although most of them have been produced with a general evolutionary outlook, such resources are included in the list because the subject matter is valuable. In addition, the teacher should be able to utilize their content as examples of the *overgeneralization* and typically *speculative thinking* that is characteristic of evolutionists. Thus, special lessons in understanding how to meet everyday expressions heard in the media can be gained.

c. Answers to most questions found at the end of chapter sections and ends of chapters of the textbook have been provided.

d. "Think-Session Guides" conclude the material for each unit. Thought-provoking questions or suggestions are interspersed in paragraphs designed to stimulate individual teacher thinking; they may be used directly (or rephrased for a particular class level) to promote class discussion prior to the commencement of each unit of study or to be used as review. Their appearance at the end of each unit in this manual is not intended to indicate that they should be used only after the student has completed the unit in the text.

Acknowledgment is hereby made of the excellent cooperative efforts of contributing authors Mrs. Olive Fischbacher, Dr. William J. Tinkle, and Dr. Ralph Paisley. Mrs. Fischbacher has participated in various science-writing projects with other publishers; Dr. William Tinkle is Emeritus Professor of Biology, Anderson College, Indiana; and Dr. Paisley has drawn upon his excellent teaching experience at The King's College, Briarcliff Manor, New York. A word of grateful appreciation is extended to these authors.

John N. Moore, Co-editor of first edition

Introduction to Teacher's Manual

A. Guidelines

1. Motivation or Enrichment

2. Multimedia Resources

3. Supplementary Reading

Two guidelines have been followed in the preparation of this teacher's manual. First, and most important, has been the safeguarding of the textbook's distinctiveness, notably the inclusion of creation in every consideration of origins. With this in view, it has seemed advisable to alert teachers regarding the suggested films and Supplementary Readings. No restrictions have been placed on such suggested materials, but asterisks and footnotes serve as reminders of possible problems in content.

The second guideline has been to afford the fullest possible assistance to home-school and classroom teachers in practical ways. These include the categories of suggestions described below.

The **Suggestions for Motivation or Enrichment** in this manual were planned to capture student interest and focus attention on the specific content of each textbook chapter. It is hoped that the suggested form of presentation—even direct use of the given wording—will facilitate utilization of these motivational concepts. Motivation is never just an end in itself, and teachers should be able to sense by student reaction just when to proceed to the succeeding item in the text sequence. Many of the suggestions also lend themselves to development as enrichment experiences and may even supplement the laboratory manual in scope.

The **Suggestions for Multimedia Resources** are representative of videos, DVDs, CD-ROMs, and Internet sources currently available. There has been no restriction placed upon the inclusion of media having an evolutionary bias, but an effort has been made to include creation-oriented resources whenever feasible. It is strongly recommended that teachers preview all material before class time and be prepared to refute any questionable statements. Many teachers will find that this has the added advantages of providing opportunity for improvised clarification of concepts and of stimulating useful dialogue.

Sources for films and laboratory supplies are listed after this introduction. Additional appropriate films may be found through local government offices, health agencies, or libraries.

The **Suggestions for Supplementary Reading** are taken largely from recent issues of periodicals available on the Internet or in public libraries. An effort has been made to prefer currently published material, but many serious students will want to check indices for additional titles.

Certain contemporary or uniquely appropriate books are also listed, with brief annotations. Such books could make valuable additions as budgets permit. Some basic references are also available at the end of some chapters in the textbook.

B. Answers to Questions

The answers to the questions are suggested solutions; some are exact answers, and some are illustrations. All of them can be elucidated or expanded by the teacher. They are intended to be guides to the teacher in his imaginative presentation of an exciting course in biological science.

C. Think-Session Guides

The teacher must read over the material and understand the biology problems involved. He should then prepare any visual materials he thinks necessary for his student or class.

After presenting the background information (if deemed necessary), he presents the problems and invites student response. Now the teacher's greatest skills come into play. He must direct the discussion by asking *diagnostic* and *analytical questions*. This procedure will help students see for themselves why certain answers are poor and others good.

Remember that getting the "right" answer is not the prime objective. Instead, the student must use his information and intelligence to search out the answer. Each answer is discussed to determine not if it is right or wrong but whether the reasoning was correct in obtaining that answer.

Resource Contact Information

Answers in Genesis

Answers in Genesis (AiG) is a Christian apologetics ministry that equips the church to uphold the authority of the Bible from the very first verse. AiG also provides biblical answers to tough questions about creation, evolution, and the Bible. Their Web site features thousands of articles covering dozens of scientific and biblical topics, plus media programs, daily devotionals, resources, and much more.

AiG's Web site also provides featured articles with a biblical perspective on current events, scientific topics, and other issues relevant to the Christian faith. Their "Question and Answer" section provides thousands of articles on creation, dinosaurs, geology, genetics, Noah's Flood, and dozens of other scientific and biblical topics. In their "Answers Media" section, you can listen online to hundreds of free audio and video programs. In addition, their "Online Bookstore" offers books, DVDs, witnessing tools, inspiring music, and much more. For curricula information visit their "Creation Education Center" Web page (www.answersingenesis.org/home/area/curriculum_info). Below is contact information for *AiG-United States*:

Mail Address: P.O. Box 510, Hebron, KY 41048
[2800 Bullittsburg Church Rd., Petersburg, KY 41080]

Phone: (859) 727-2222

Customer service: (800) 778-3390

Ministry information: (800) 350-3232

Web site: www.answersingenesis.org

E-mail: Use their *www.answersingenesis.org/feedback* page for specific questions.

Creation Research Society, The

The Creation Research Society (CRS) is a professional organization of trained scientists and interested laypersons who are firmly committed to scientific *special creation*. The Society was organized in 1963 by a committee of ten like-minded scientists and has grown into an organization with worldwide membership. The primary functions of the Society are (1) the publication of a quarterly peer-reviewed journal (*CRS Quarterly*), (2) the conducting of research to develop and test creation models, and (3) the provision of research grants and facilities to creation scientists for approved research projects. The *Creation Research Society Quarterly* presents scientific evidence supporting intelligent design, a recent creation (young earth creationism), and a catastrophic worldwide flood. The CRS produced the first edition of this textbook in the early 1970s. Their online bookstore carries a large selection of creation books, videos, and DVDs.

Address: P.O. Box 8263, St. Joseph, MO 64508-8263

Web site: www.creationresearch.org

E-mail: contact@creationresearch.org

The CRS has also established the Van Andel Creation Research Center in north-central Arizona for the purpose of aiding the Society and other visiting scientists in their research efforts. The Society encourages a broad spectrum of research to develop and test a *creation model* and administers a research grant program whereby modest funds are distributed to qualified researchers for the conduct of creation-related research.

For additional information not provided on their Web site about the CRS Van Andel Research Center, contact Dr. Kevin Anderson:

Address: 6801 N Highway 89, Chino Valley, AZ 86323-9186

Phone: (928) 636-1153

E-mail: vacrc@creationresearch.org

Fogware Publishing

Fogware Publishing provides educational media for grades K–12.

Address: 9625 West 76th Street, Suite 150, Eden Prairie, MN 55344

Phone: (408) 977-0250

Web site: www.fogwarepublishing.com

E-mail: support@jcresearch.com

Illustra Media

"Now a superb new documentary from Illustra Media, entitled *Unlocking the Mystery of Life*, explains how biological evidence both challenges Darwinian evolution and strongly supports the alternative theory of intelligent design."

Charles Colson, *Breakpoint*, May 21, 2002

Address: 18005 Sky Park Circle, Suite K, Irvine, CA 92614

Phone: (800) 266-7741 or (949) 794-9109

Web site: www.illustramedia.com

E-mail: felicia@go2rpi.com

Institute for Creation Research

The Institute for Creation Research (ICR) offers seminars, conferences, debates, various other speaking engagements, creation science workshops, radio and television outreach, creation research, guided tours to areas of geological interest, books, videos, publications, and free periodicals. The ICR also houses the Museum of Creation and Earth History, ICR Graduate School, library, science labs (one with an electron microscope), computer center, art center, radio broadcast facility, as well as various other facilities.

Address: 10946 Woodside Ave., North, Santee, CA 92071

Phone: (619) 448-0900

Fax: (619) 448-3469

Web site: www.icr.org

Moody Publishers

Moody Video, a division of Moody Publishers, has exciting videos/DVDs that bridge the gap between science and faith. They have a wide selection of items for family entertainment, faith and reflection, Christian education, and homeschooling. Both educational and fun to watch, Moody Videos/DVDs show us that faith can be combined with science, helping us to understand the majesty of God through the wonders of His creation.

Address: 820 N. La Salle Blvd., Chicago, IL 60610

Phone: (800) 678-6928 or (800) 842-1223 or (312) 329-2101

Web site: www.moodyvideo.org *or* www.moodypublishers.com

E-mail: mpcustomerservice@moody.edu *or* pressinfo@moody.edu

Moody Science Classics, 20 DVD Set[1]

Marvel at amazing animals. Journey through the human circulatory system. Discover why the Incan culture mysteriously declined.... And through it all, rejoice in the majesty of God! This set of award-winning DVDs offers your whole family a balanced approach to science, while giving credit to the Creator. Addresses physics, biology, astronomy, and more! 28 minutes each.

City of the Bees, Dust or Destiny, Empty Cities, Facts of Faith, God of Creation, Hidden Treasures, Prior Claim, Red River of Life, Signposts Aloft, Where the Waters Run, Windows of the Soul, God of the Atom, Of Books and Sloths, Professor and the Prophets, Ultimate Adventure, Experience with an Eel, Mystery of the Three Clocks, Time and Eternity, Voice of the Deep, Journey of Life

NASA Central Operation of Resources for Educators[2]

The NASA Central Operation of Resources for Educators (CORE), established in cooperation with Lorain County Joint Vocational School, serves as the worldwide distribution center for NASA-produced multimedia materials. For a minimal charge, CORE will provide a valuable service to educators unable to visit one of the NASA Educator Resource Centers by making NASA educational materials available through its mail order service.

Through CORE's distribution network, the public has access to more than 200 video, DVD, slide, and CD-ROM programs, chronicling NASA's state-of-the-art research and technology. Through the use of these curriculum supplement materials, teachers can provide their students with the latest in aerospace information. NASA's educational materials on aeronautics and space provide a springboard for classroom discussion of life science, physical science, space science, energy, earth science, mathematics, technology, and career education.

Mail Address: Central Operation of Resources for Educators (CORE)
Lorain County Joint Vocational School
15181 Route 58 South, Oberlin, OH 44074

Phone: (866) 776-CORE (*or* 2673) *or* (440) 775-1400

Fax: (866) 775-1460 or (440) 775-1460

Web site: core.nasa.gov *or* education.nasa.gov/edprograms/core/home/index.html

E-mail: nasa_order@leeca.org

National Agricultural Library, The

The National Agricultural Library (NAL) supplies agricultural materials to other libraries and information centers. You should submit your requests through your local library. In the United States, possible sources are public libraries, state libraries, land-grant university libraries, or other large research libraries in your state. If the publications are not available, have the library submit an *interlibrary loan request* to the NAL. Look for the "NAL CALL Number" or search online at <agricola.nal.usda.gov>.

---

1. This 20 DVD set is available at a discount from **www.christianbook.com**; search for item number 679658. For phone orders, call 800-247-4784. Most videos in this series are still available at a discount, but they are going out of print.

2. **NASA CORE** is a service of the *Education Division* of the *National Aeronautics and Space Administration* and *Lorain County Joint Vocational School.*

Outside of North America, individuals should request materials through major university, national, or provincial institutions. Requestors in countries with an AGLINET (Agricultural Libraries Network) library are encouraged to make full use of that library and its networking capabilities. To access the AGLINET Web site, go to <www.fao.org/library> and choose "Partnerships-Networks."

Address: National Agricultural Library, Abraham Lincoln Building
10301 Baltimore Avenue, Beltsville, MD 20705-2351

Phone: (301) 504-5755

Web site: agricola.nal.usda.gov

Phoenix Learning Group, Inc., The
Divisions: Phoenix Films & Video, BFA Educational Media, Phoenix Learning Resources, and Coronet/MTI

In 1973, Heinz Gelles and Barbara Bryant formed a company called *Phoenix Films & Video*, committed to producing quality educational films. After Phoenix Films became established, *BFA Educational Media* was purchased from CBS. This division is geared mainly toward the educational marketplace.

In 1993, Phoenix moved its corporate headquarters from New York City to St. Louis, Missouri. In conjunction with this relocation to St. Louis, all company divisions were organized as a division of *The Phoenix Learning Group, Inc.*

In 1997, Phoenix acquired a fourth division—*Coronet/MTI Films* and its award-winning collection of educational videos, videodiscs, and multimedia programs—from Simon & Schuster, the publishing operation of Viacom, Inc.

Approximately 6,000 titles are now under the company's ownership or control. Phoenix currently provides quality educational content for almost any technology now developed. These titles are available in a variety of formats: 16 mm, video, laser-disc, CD-ROM, and DVD.

Address: 2349 Chaffee Drive, Saint Louis, MO 63146

Phone: (314) 569-0211or (800) 221-1274 to request their Home School catalog

Fax: (314) 569-2834

Web site: www.phoenixlearninggroup.com (to request an age-specific catalog online)

E-mail: multimediasales@phoenixlearninggroup.com
(contact: Erin Bryant, Vice President of Operations and Management)

Bio Corporation

Bio Corporation offers quality educational materials at reasonable prices. All the preserved specimens, live specimens, dissection equipment, educational videos, educational CDs, anatomical charts, and 3D models that you are looking for are available.

Address: 3911 Nevada Street, Alexandria, MN 56308

Phone: (800) 222-9094 to request their Home School catalog

Fax: (800) 332-9094

Web site: biologyproducts.com (to request a catalog online)

E-mail: online or at <biocorp@rea-alp.com>

Scientific Equipment Resources

Bob Jones University Press

Bob Jones University Press has a detailed, step-by-step Biology Dissection Labs video guide on the dissection of a crayfish, perch, earthworm, and frog. This video is available in both DVD and VHS formats. This would be a very helpful product for students who are studying chapters 14 and 15. For further information about this product, simply go to the Bob Jones Web site and search for "biology dissection labs."

Address: BJU Press, Customer Services, Greenville, SC 29614-0062

Phone: (800) 845-5731 *or* (864) 242-5100, ext.3300

Fax: (800) 525-8398 *or* (864) 271-8151 *(orders only)*

Web site: www.bjup.com

E-mail: bjupinfo@bjup.com

Carolina Biological Supply Co.

Carolina Biological Supply Company has a wide selection of scientific equipment, including preserved specimens for use in dissections. The company has established a page on its Web site that is specifically designed for the home education community, featuring resources that the company believes are particularly appropriate for home schoolers.

Address: 2700 York Road, Burlington, NC 27215-3398

Phone: (800) 334-5551 (US), *or* (800) 933-7833 *or* (519) 737-9212 (Canada)

Fax: (800) 222-7112 (US), *or* (519) 737-7901 (Canada)

Web site: www.carolina.com

Home School Web page: www.carolina.com/homeschool/

E-mail: carolina@carolina.com

Edmund Scientifics

Address: 60 Pearce Avenue, Tonawanda, NY 14150-6711

Phone: (800) 728-6999

Fax: (800) 828-3299

Web site: www.scientificsonline.com

E-mail: scientifics@edsci.com

Flinn Scientific, Inc.

Flinn Scientific has a helpful Web site for biology teachers, with suggested activities, useful information, and links to biology-oriented Web sites.

Address: P.O. Box 219, Batavia, IL 60510

Phone: (800) 452-1261

Fax: (866) 452-1436

Web site: www.flinnsci.com

E-mail: flinn@flinnsci.com

Home Training Tools

Home Training Tools has been in business for ten years, helping families to understand and delight in God's creation through science. This site is owned and operated by Frank and Debbie Schaner. They share a commitment with families who are seeking effective ways to teach science at home. Home Training Tools is truly a family business that also involves their four children, who have always been homeschooled.

Address: 665 Carbon Street, Billings, MT 59102

Phone: (800) 860-6272 or (406) 256-0990

Fax: (888) 860-2344 or (406) 256-0991

Web site: www.hometrainingtools.com

E-mail: service@hometrainingtools.com

NASCO

NASCO has a large number of materials that can be used for dissection activities. These include both biological specimens and dissection alternatives.

Address: 901 Janesville Avenue, P.O. Box 901, Ft. Atkinson, WI 53538
4825 Stoddard Road, P.O. Box 3837, Modesto, CA 95352-3837

Phone: (800) 558-9595 *or* (920) 563-2446

Fax: (920) 563-8296; Modesto, CA (209) 545-1669

Web site: www.enasco.com

E-mail: custserv@enasco.com

Nebraska Scientific

Address: 3823 Leavenworth Street, Omaha, NE 68105-1180

Phone: (800) 228-7117; **Fax:** (402) 346-2216

Web site: www.nebraskascientific.com

E-mail: staff@nebraskascientific.com

Schoolmasters Science

Address: 745 State Circle, Box 1941, Ann Arbor, MI 48106

Phone: (800) 521-2832; **Fax:** (800) 654-4321

Web site: www.schoolmasters.com

E-mail: online or at <wolverine@school-tech.com>

WARD'S Natural Science

Address: 5100 West Henrietta Road, Rochester, NY 14692-9012
812 Fiero Lane, P.O. Box 5010, San Luis Obispo, CA 93403-5010

Phone: (800) 962-2660 or (585) 359-2502; San Luis Obispo, CA (800) 872-7289

Fax: (585) 334-6174; San Luis Obispo, CA (805) 781-2704

Web site: wardsci.com

E-mail: customer_service@wardsci.com

UNIT 1
Science: Finding Order in Complexity

CHAPTER 1

The Scientist and His Methods
Text Pages 3–12

Work of the Scientist
Goals of the Scientist
Specialization of the Scientist

◆ Suggestions for Motivation or Enrichment:[*]

1. As you think of some common "-ologies" which describe fields of science, like *biology* and *zoology*, make up a few "-ologies" for yourself, even if you have never considered them before. Include your own words for the study of mammals and for the study of viruses. Do not hesitate to originate a word, providing you can justify your decision. Then with a dictionary or other resource aid, discover if there is a better term than you have chosen in each case.

2. Choose an outstanding scientist, such as Einstein or Edison, who has been widely described in literature, and research a list of his specific accomplishments, noting which ones seem to have been the result of incidental discovery, and which ones seem to have been planned as such from the beginning of his work.

3. Think of some incidental information which a soil bacteriologist might find out as he goes about his study of bacteria in soil, and write out a hypothetical report of his incidental findings as if you had worked with him. Think in terms of such things as soil acidity, pollutants in the area, and plant growth.

4. Presume that you have the responsibility of hiring a drinking water analyst for a city laboratory, and state the qualifications that you consider vital for such a position. Make your statements in writing as if they represent a group of applicants for the position.

◆ Suggestions for Multimedia Resources:[*]

1. *Creation* (Answers in Genesis, CD-ROM). Answers in Genesis wants to do all it can to counter the massive propaganda push by PBS, NOVA, and Clear Blue Sky productions. Millions of dollars in funding were spent in the production of seven anti-Christian programs in the television series entitled *Evolution*. As a response to this series, and to equip people to defend the Christian faith, Answers in Genesis has created this multimedia-rich CD-ROM.

2. *Project 3:15* (Answers in Genesis, CD-ROM). With hundreds of pages of articles, over 10 hours of audio files, and the complete video *The Image of God,* this CD will effectively equip you to answer many timely questions. It also contains Dr. Jonathan Sarfati's entire book *Refuting Evolution 2.*

3. The "Riddle" of Origins, video/DVDs: *Does it Matter What We Believe?, The Origin of Life, The Origin of Humans,* and *Dating Fossils and Rocks* (Institute for Creation Research, approximately 60 min. each). Mike Riddle,[**] former Olympic athlete and Microsoft communications expert, emboldens students of all ages to defend their faith in the Bible and become more effective witnesses.

4. *Episode 4: All Systems Go* (NASA CORE, 22 min.). This videotape shows research conducted aboard the space shuttle on six systems that examine the heart, lungs, blood, muscles, cells, and the immune system, among others.

5. *Science in Space: Fundamental Biology on STS-107* (NASA CORE, 34 min.). This video on the space shuttle Columbia STS-107 mission (2003) is dedicated to research investigating human physiology, fire suppression, and other areas of research relevant to people across the globe.[***]

6. *Unlocking the Mystery of Life*/DVD and Video (Illustra Media, 67 min.). This presentation explores timeless questions on the origins of life and offers compelling evidence to support an idea that could revolutionize scientific thought—the theory of intelligent design. (Also available: *The Privileged Planet* and *Where Does the Evidence Lead?*)

7. *God of Creation*/DVD and Video (Moody, 28 min.). This all-time Moody favorite explores the power and beauty of the universe we live in. You will see the food "factories" that plants use and peer through a telescope into the vast reaches of space, while you ponder the God who put the galaxies in place—the same God who gave His Life for us on the cross.

---

[*] The listing of these suggestions does not necessarily imply endorsement of content.

[**] Mike is the president of **Christian Training and Development Services** (http://www.train2equip.com/index.asp) and currently travels around the country teaching and delivering seminars on the biblical and scientific truths of creation.

[***] To see online **Tribute Videos**, visit <spaceflight.nasa.gov/shuttle/archives/sts-107/memorial>.

◆ **Suggestions for Supplementary Reading:** [*]

1. Behe, M.J., W.A. Dembski, and S.C. Meyer. October 2000. *Science and Evidence for Design in the Universe*. Ignatius Press, 234 pages. In a lucid primer to the Intelligent Design debate, three neo-creationists summarize their most important data and conclusions and rebut critics. It examines how the complexity and interrelatedness of creation argue for a Creator.

2. Bevington, Linda K., *et al. Basic Questions on Genetics, Stem Cell Research, and Cloning: Are These Technologies Okay to Use?* BioBasics Series. Kregel Publications, 125 pages. This booklet uses a biblical perspective to examine the scientific and ethical issues surrounding such interventions. It not only clarifies current and future developments but also gives guidelines for developing a Christian response.

3. Marsch, Glenn A. "The Book of the Law and the Book of Nature: A Two-Volume Set?" *Scientific Voice, Center for Scientific Studies*. To read this article online, visit Dr. Marsch's own Web site at <http://members.aol.com/drgmarsch/faithframes.html>. It discusses the relationship between general and special revelation for the Christian scientist.

4. Mortenson, Terry. November 2004. "*National Geographic* is wrong and so was Darwin." *Answers in Genesis*. To read this article online, visit <http://www.answersingenesis.org/docs2004/1106ng.asp>. It refutes the 33-page cover story of the November 2004 issue of *National Geographic* (*NG*). Scientifically informed and careful thinking readers will want to analyze *NG*'s "overwhelming evidence" before accepting its conclusion.

5. Sarfati, Jonathan. June 1999. *Refuting Evolution*. Master Books, 176 pages. This book is a general critique of the most up-to-date arguments for evolution; it will challenge educators, students, and parents alike.

◆ **Answers to Questions**

Work of the Scientist (Text page 9)

➤ **Questions: Work of the Scientist**

1. The term *science* is the study of God's creation. Because creation was brought into existence by God's wisdom, and man was created as a part of it, man is called to fulfill the Creation Mandate—to understand and subdue creation for God's glory.

2. The assumption of *uniformity*—that nature is orderly and that events can be repeated—is fundamental to the study of science.

3. A *hypothesis* is only a temporary explanation formed to explain a problem, and *it must be testable*. However, a *theory* involves a broad range of concepts concerning many problems and usually includes some imagined aspect; thus, *no theory can be tested directly*.

4. A *law* is more extensive than a *theory* because it has been well established by the efforts of many researchers—who repeat the steps, reach the same conclusion, and rule out other explanations.

5. The criteria of a good theory are (1) to identify the *orderly relationship* of many seemingly diverse and isolated observations of the natural environment; (2) to *predict* certain future events; these predictions become an indirect means of confirming a theory because they are limited in scope and they are subjected to testing, as a hypothesis; (3) to be *modifiable* so that adjustments can be made as new data are accumulated, or as ideas about the imagined unit or aspect need to be changed; and (4) to *develop new directions* for research so that new observations of the natural environment can be collected.

6. It is important to repeat experiments to discover an orderly sequence of events, usually assuming that the first event is the cause of the second.

7. A *control group* is needed as a standard of comparison in judging the effects of an experiment in which the subjects are treated the same except for the omission of the procedure or agent under test, which the *experimental group* receives.

➤ **Taking it Further: Work of the Scientist**

1. It is impossible to prove scientifically how the earth was formed because it was an unrepeatable event that took place in the past; it cannot be tested.

2. It is possible to test for more than one variable at a time, but it would be impossible to determine accurately which variable caused the results that were observed. Yet, multiple variables in combination may cause the desired effect observed.

3. Science has the self-imposed limitation of studying only those things that can be demonstrated by or to the senses. Science is also limited by the ability and objectivity of those who do the research. Science is also influenced by prevailing views of the time, which may limit the scope of other scientific endeavors. The inability to experiment also imposes limitations in such fields as astronomy or archaeology. Moreover, experimentation is limited in time and space.

Goals of the Scientist (Text page 10)

➤ **Questions: Goals of the Scientist**

1. It is necessary for the findings of a scientist to be checked by other scientists before being accepted

[*] The listing of these suggestions does not necessarily imply endorsement of content.

because the scientist may have developed an erroneous hypothesis, overlooked certain factors or problems, "observed" something he wanted to see, or been sloppy in his methodology.

2. *Supernaturalism* is the belief in a supernatural being or power responsible for the created order and that intervenes in the course of natural laws.

3. "Pure science" is knowledge pursued for its own sake, and "applied science" is knowledge pursued for the material benefit of man.

➤ **Taking it Further: Goals of the Scientist**

1. The teleological views of the Bible transformed the idea of *cause and effect*, based on the Word of God. There is not only a pattern but also a plan in the mind of the Creator, which gives purpose to all of creation. Greek mythology, however, taught that the universe was constructed according to a pattern that could only be demonstrated through a study of cause-and-effect relationships.

Creationists believe that creation is the sum total of acts by God the Creator, who brought the universe into existence. Evolutionists, however, believe that all life came from an inorganic beginning and that human life came from one-celled forms through multicellular organizations of two-cell layered and three-cell layered forms of animals.

2. While the findings of science are not absolutely supreme, it is in the application of such findings that the questions of morality are introduced; for example, the sacrifice of the unborn in stem cell research, supposedly for the good of others. Such issues as cloning, genetic engineering, and stem cell research raise many moral questions that need to be addressed.*

Specialization of the Scientist (Text page 12)
➤ **Questions: Specialization of the Scientist**

1. Biology in general is the science of *life*, which is the subject matter included in biology but not in the physical sciences.

2. The five main characteristics of living things are as follows: (1) *they respond to stimuli*, (2) *they are highly organized*, (3) *they metabolize*, (4) *they grow*, and (5) *they reproduce*.

3. Biology has contributed in the field of agriculture by increasing the productivity of the soil, developing disease-resistant varieties of plants and animals, and combating plant and animal diseases when they arise. In the field of medicine, biology has helped to increase life expectancy. Biologists have also contributed substantially in such fields as conservation, fisheries, and public health.

---

* For more information, visit the Web site for *The Center for Bioethics and Human Dignity* at <http://www.cbhd.org/>.

➤ **Taking It Further: Specialization of the Scientist**

1. The difference between nonliving material and living material is that *living systems show all five characteristics*, whereas *a single nonliving system shows at most one or two of these characteristics*, and only in a manner unrelated to its welfare or continuation. Therefore, it is possible to distinguish between them by determining if all five characteristics are present at once or not.

➤ **Questions: Chapter Review**

1. *Science* is the study of God's creation. The physical sciences have benefited mankind with the discovery of each new energy source (coal, petroleum, natural gas, atomic power, solar energy, etc.), which brings greater wealth and less physical toil. The biological sciences have also helped to grow better food crops to support more and more people. Developments in medicine have been used to substantially lengthen the average life span. Many dread diseases (smallpox, bubonic plague, polio, and tuberculosis) have been conquered; and new surgical techniques have also corrected conditions that at one time were hopeless.

2. Greeks approached science by reasoning without making observations; they despised experimentation. Today, scientists use the scientific method to establish a valid interpretation of facts that has not previously been known; they are not satisfied with mere data or facts, which consist of observations and measurements, but desire to find out the reason for the facts being what they are.

3. The work of a scientist consists of stating a **problem** that no one else has studied or solved. The scientist gathers many facts, or does research, which may have a bearing on the problem. A **hypothesis**, or estimate, is formed that might explain the problem. More facts are gathered, and their relevance to the hypothesis is carefully weighed. If possible, experiments are performed. If the facts gathered are consistent with the suggested explanation, or hypothesis, the scientist concludes that the explanation is valid, and the results are published.

Scientists have agreed to deal only with *sense observations* and to limit themselves to a study of those things that can be demonstrated by or to the senses. They are also limited by those phenomena that can be repeated. Often they are limited by their own limited knowledge and trends of the day.

4. The assertion, "scientists pride themselves on their objectivity," means they try to look only at the evidence and draw conclusions only supported by the evidence; in weighing facts, they are presumably impartial and unemotional. Yet, they have an understandable emotional attachment to the concepts they have developed or embraced and are

influenced by prevailing views of their time. *Examples may vary.*

5. The primary goal of the scientist is *objectivity*, but he also seeks to find *purpose* in the created order—to discover not only a pattern but also a plan in the mind of the Creator. A third goal of the scientist is to discover *truths* regarding the natural world, recognizing that all scientific conclusions are tentative (only the Bible provides absolute truth).

6. A hypothesis is important as a working basis; it presents something which may be true, to be tested by logic or experiment.

7. Scientists may rule out facts that point toward a conclusion that is "undesired," judging that those facts have no bearing on their hypotheses or their theory of origins. However, such *undesired conclusions* should not be overlooked. Also, scientists may not know how present scientific "truths" or concepts will be changed, but we can be certain that they will; they are considered to be *tentative conclusions*. Finally, *correct conclusions* are unmixed with error; they require a disciplined mind, guided by the principles of God's Word.

8. Since we live in a universe of absolutes with absolute time and absolute space, the role of science is to discover **absolute truth**. Moreover, the scientist would not be able to function or have a purpose in a world without absolutes.

9. It is difficult to define *life* from a scientific perspective because life is unique—there is nothing like it. Yet, for an adequate definition, the subject (e.g., of life) must be placed into a broad category and then described as to how it differs from other subjects in the category. However, there is no broad category where the subject of life can be placed.

Answers may vary. The student is also asked to explain how would a creationist or evolutionist define life. According to evolutionists, new species are created by a purposeless, random process of genetic mutation. However, if you study the biological world with an open mind, you will see more evidence that each separate species was created by an Intelligent Designer (i.e, God). Moreover, evolutionists have not found the fossils of any transitional species—half reptile and half bird, for instance. Similarly, there are no rich fossil deposits before the Cambrian era (550 million years ago). If Darwin was right, what happened to the fossils of all their evolutionary predecessors? Evolution is unscientific, because it is not testable or falsifiable; it makes claims about events (such as the very beginning of life on earth) that can never be recreated. A better solution is that God, or a designer, deposited each new species on the planet, fully formed and marked "made in heaven."[*]

CHAPTER 2
Application of Scientific Methods to the Insect World
Text Pages 15-24

Knowledge of Insects
Management of Insects
Description of Common Insects

◆ **Suggestions for Motivation or Enrichment:**

1. Make a list of the insects that you personally dislike most and give reasons for your choices. Make another list of insects that may be serving a good purpose in your life and tell how they serve.

2. Prepare a report on a present-day insect problem as it is related to world travel, ecology, housing, disease, or food supplies.

3. Write out a practical plan for organic gardening, identifying the relationships it would have to insect life. You may think you know who your insect "friends" are, but beware: *Your garden could be crawling with impostors that are actually pests.*

4. Presume that you are directing some young scientists to study a strange new insect for classification purposes; as their director, present your guidelines or instructions to your colleagues.

◆ **Suggestions for Multimedia Resources:**[**]

1. *City of the Bees*/DVD and Video (Moody, 28 min.). Examine the complex community of the bees and marvel at their amazing sophistication. Learn why God's design for humans is vastly superior.

2. *Pests In and Around the Home*/CD-ROM (University of Florida/IFAS). This CD contains a computerized knowledge base of house pests, plus information on pest biology, life cycle, identification, distribution, damage, and management. Hundreds of scientific definitions, graphics, and photos are also included. (For more information, visit <http://pests.ifas.ufl.edu/software/det_pests.htm>.)[***]

3. The Department of Entomology at Michigan State University works with farmers and homeowners to better understand how insects affect our lives and how to effectively manage our interactions with them. Their Web site has many helpful articles. (For more information, visit <www.ent.msu.edu/Extension/extension.htm>.)

---

[*] Visit <www.wasdarwinright.net> for more information.

[**] The listing of these suggestions does not necessarily imply endorsement of content.

[***] For information on other insect software, see the UF/IFAS **Buggy Software** site at <http://pests.ifas.ufl.edu/software>.

4. *Biological Control: A Guide to Natural Enemies in North America* (Cornell University). To read this guide, visit <http://www.nysaes.cornell.edu/ent/biocontrol/>). This site provides photographs and descriptions of biological control agents of insect, disease, and weed pests in North America. It is also a tutorial on the concept and practice of biological control and integrated pest management (IPM).

◆ **Suggestions for Supplementary Reading:**[*]

1. Hoffmann, M. P., and A. C. Frodsham. *Natural Enemies of Vegetable Insect Pests.* Cooperative Extension, Cornell University, Ithaca, NY. 1993, 63 pp.

2. Zahl, Paul A. "What's So Special About Spiders?" *National Geographic* (August 1971), 190–219. To view this article online, visit the following Web page: <http://staffwww.fullcoll.edu/lvincent/ZAHL-190.htm>. (To turn to the next page of the article, use the "next" button at the bottom of the page.)[**]

3. Moffett, Mark W. "Big Bite" *National Geographic* (July 2004). Massive jaws, voracious appetite, and sprinters' speed attest that these aggressive desert dwellers (wind scorpions) are built to kill.

◆ **Answers to Questions**

Knowledge of Insects (Text page 16)

➢ **Questions: Knowledge of Insects**

1. Entomology, or the study of insects, is the division of biology that is the subject of this chapter.

2. A *niche* is the ecological role of an organism in a community, especially in regard to food consumption. The two main classes of niches include **producers** (green plants) and **consumers** (animals).

3. The different types of consumers are *herbivores, carnivores,* and *omnivores.* Also, carnivores (either plant or animal) may be classified as *predators* or *parasites.* Some organisms are *scavengers* that consume dead plant and animal material. *Examples of each from the insect kingdom will vary.*

➢ **Taking it Further: Knowledge of Insects**

1. Definitions for and the Greek or Latin root of:

   • **Ecology:** study of the interrelationship of organisms and their environments; Greek: *oikos,* "home"; *logos,* "study of"
   • **Parasites:** organisms that live on or within a living host and cause it harm; Greek: *parasitos < para-,* "beside, alongside of"; *sitos,* "grain, food"
   • **Herbivores:** plant-eating animals; Latin: *herba,* "plant"; *vore,* "eat"

• **Carnivores:** flesh-eating animals; Latin: *carne,* "flesh"; *vore,* "eat"
• **Omnivores:** organisms that eat both plants and animals; Latin: *omni,* "all"; *vore,* "eat"

2. With so much destruction and disease caused by insects, man has not eliminated them altogether because controlling these insects is a complex problem. The farmer, regardless of the plants grown, is faced with the constant threat of a "population explosion" among the ever present herbivorous insects.

Management of Insects (Text page 21)

➢ **Questions: Management of Insects**

1. Methods of controlling insects:

   • **Insecticides** are used more than any other method of insect control.
   • **Environmental controls** require more detailed knowledge of the life history of insects and where they spend each stage of their life.
   • The ideal way to control insects is to develop and grow **insect-resistant plants**.
   • **Biological controls** are used to reduce the population of many insect pests by introducing their natural enemies—carnivores (predators and parasites) that prey on them.

2. Disadvantages of insecticides:

   • poisonous to insecticide handlers or to those coming in contact with them
   • may remain on food plants, such as apples or lettuce, and become a health hazard
   • difficult to control, due to the wind, rain, etc.
   • insect populations "develop" resistance to insecticides, passing on resistance genetically to offspring
   • may kill desirable organisms; e.g., honeybees are killed by indiscriminate applications of DDT

3. An example of biologic control is the Vedalia beetle, which was used to control cottony-cushion scale insects that attacked citrus orchards in California in 1887. *Other examples are possible.*

➢ **Taking It Further: Management of Insects**

1. *Answers may vary. Internet or library resources may be used to answer this problem.*

   The *Anopheles* mosquito is an insect that carries the dread disease malaria, now devastating Sub-Saharan Africa. DDT (*dichlorodiphenyltrichloroethane*) is a well-known insecticide that is effective in controlling the species of *Anopheles* that carry malaria. Placing live mosquito fish (*Gambusia affinis*) in ditches and ponds to eat mosquito larvae is one possible biological control.[***]

---

[*] The listing of these suggestions does not necessarily imply endorsement of content.

[**] Other related articles may be found at the following Web page: <http://staffwww.fullcoll.edu/lvincent/READINGS.htm>. Just click on a particular author to access the article. Note that, most likely, the authors on this Web site adhere to an evolutionary worldview.

[***] Other non-chemical control methods include invertebrate predators, parasites, and diseases to control *mosquito larvae.* Birds, bats, dragonflies, and frogs may control *adult mosquitoes;* however, supportive data is anecdotal, and there is no documented study to show that such predators consume enough adult mosquitoes to be effective control agents.

2. An understanding of entomology would be useful in the following career fields: agriculture, forestry, gardening, health care, or any field that has to due with living organisms. In the case of agriculture, there is a need to control insects that destroy crops or that are harmful to livestock. In the case of health care, the control of mosquitoes is critical in the fight against malaria, West Nile Virus, etc.

Description of Common Insects (Text page 24)

➤ **Questions: Description of Common Insects**

1. *Answers will vary.*

2. *Answers will vary.*

   The common housefly (*Musca domestica*) may be controlled by the removal of their breeding places (such as manure and garbage), by killing the few individuals that live through winter, and by using screens to keep them out of houses; also, some insecticides can be useful.

3. Houseflies are so dangerous because they cause more deaths per year than all other animals combined. Germs not only cling to their feet but also enter their digestive tracts; these germs are then spread through their waste and saliva.

➤ **Taking It Further: Description of Common Insects**

1. The difficulty encountered by the notion of clover being on the earth thousands of years before bees is that clover requires bees to pollinate its flowers. If clover flowers are isolated from insects, the flowers produce very few seeds.

2. The terms *carnivore* and *predator* do overlap, since carnivores that kill prey and eat it immediately are called predators.

➤ **Questions: Review**

1. It important to have a balance of the different types of consumers because so many young would be produced that they would soon eliminate their own food supply if not checked.

2. Classification of any organism is arbitrary; for example, the structure of insects is usually the basis for their classification. But there is no reason why they could not be classified on the basis of where they live. So, grouping together all those that live in the ground, those that live on the ground, and those that spend most of the time in the air would be possible. The possibilities are endless.

3. An *exotic species* is one that is not in its native habitat. It was carried to another land where it did not have natural enemies, which in most cases were carnivorous insects. Many of the most harmful insect pests are exotic species.

4. *Answers will vary. Since this is not answered directly in the text, supplementary material may be used.*

Examples of natural parasites of insect pests are as follows:

- parasites used to control the fern weevil in Hawaii
- parasites (parasitic wasps) are used to control the sugar beet leafhopper[*]

5. The steps in the biologic control of an insect are as follows:

- From the form of the insect, learn its name and observe what other scientists have learned about it.
- Learn the area from which the harmful insect came.
- Go to that area and look for an enemy of the insect.
- Introduce the enemy of the harmful insect into the area where the harmful insect now lives.

6. Some pitfalls for each step in question 5 are:

- Duplicating the research that has already been done serves no useful purpose.
- If the insect pest has been misidentified, the wrong area may be searched in vain.
- The enemy, however, may have its own enemies (*hyperparasites*); thus, it should not be obtained.
- There is the danger of introducing other harmful species of insects.

7. Definitions for and the Greek or Latin root of:

- **Hyperparasites:** parasites that are parasitic upon other parasites; Greek: *hyper*, "above, beyond"; *para-*, "beside, alongside of"; *sitos*, "grain, food"
- **Pathogens:** specific causative agents (bacteria, viruses, or other microorganisms) of disease; Greek: *pathos*, "suffering"; *genos*, "birth"

8. Several insects commonly found in the home, garden, and farm are as follows:

- **Home:** houseflies, cockroaches, clothes moths, carpet beetles, carpenter ants, etc.
- **Garden:** aphids, ladybugs, lacewings, hover flies, scale insects, mealy bugs, green stink bugs, squash bugs, lace bugs, ants, butterflies (e.g., cabbage white), moths (e.g., Carolina sphinx), etc.
- **Farm:** horn flies, stable flies, horse flies, screwworm flies, heel flies, etc.

9. The means of controlling the insects in each of the above locations are as follows:

- **Home:** removal of their breeding places, killing the few individuals that live through winter, use of screens to keep pests out, some insecticides, dust and sprays, dry cleaning and storage, fumigants
- **Garden:** ladybugs (lady beetles or ladybirds), lacewings (lacewing flies), and the larvae of hover flies eat aphids; insecticides and sprays may be used
- **Farm:** insecticides (not common), sterilized (by irradiation) male insects released to mate

---

[*] Sugar beet leafhoppers do very little direct damage to the crop but are a serious pest because they vector the curly top virus.

◆ Think-Session Guide for Unit 1

Finding Order in Complexity

Subject: Seed Germination

Purpose: To Learn How a Scientist Thinks

Teacher: In his "search for order" the student may misinterpret data that he has gathered personally or has read in a scientific journal. Thus, the teacher must guide each step as the student learns how to evaluate and interpret data and design experiments.

---

a. **To the student:** An experiment was performed to consider the conditions necessary for *seed germination*. Several seeds were placed on moist filter paper in each of two glass *petri* (pē′trē) *dishes*. One of the dishes was placed in the dark; the other was placed in continuous light. Both were kept at the same temperature. After four days, the dishes were examined. It was found that all the seeds in both dishes had germinated.

What interpretation would you make of the data presented?

Teacher: *Light* is the factor being tested here. However, a student may suggest that moisture and/or warm temperature is also necessary. If these factors are not mentioned, introduce one as a possibility.
If such interpretations do come, suggest that the glass dish is also necessary. The student will probably not accept this. At this point you should have little difficulty showing that (1) not all data are usable, (2) one must be careful in interpreting data, and (3) data properly interpreted provides evidence for conclusions.

---

b. **To the student:** What condition was obviously different between the two dishes? Since this was a planned investigation, state precisely the *problem* that precluded the particular plan of the experiment.

Teacher: At this point, the student should realize that the experimenter had in mind the general problem: "What environmental conditions control seed germination?" This question is too general to lead to an experimental design and cannot be answered easily. The specific problem, "Do seeds need light to germinate?" was immediately solvable by experimentation.

---

c. **To the student:** Considering the problem you have just stated, look at the data again. What conclusion can be drawn?

Teacher: The evidence should now indicate that light is not necessary for the germination of *some* seeds. (Point out that Grand Rapids lettuce is stimulated to germinate in light, whereas Great Lakes lettuce is inhibited by light.)

---

d. **To the student:** Although "light" is not *necessary* for the germination of these seeds, different "amounts of light" may play some role. How could the investigator examine this question?

Teacher: Count the number of germinated seeds each day in both conditions. Different light intensities and/or colors may also affect the results.

---

e. **To the student:** By now you should have some "feel" for *how* a scientist works. Design an experiment in which you test the "effect of temperature" on seed germination.

Teacher: The experiment should have two groups of seeds under the same moisture and light conditions but varying in temperature.
The concepts of *hypothesis* and *control* can be introduced here.

UNIT 2
Chemical Perspectives in Biology

CHAPTER 3
Basic Chemical Principles
Text pages 27-34

Inside the Atom

Molecular Structure

Inorganic and Organic Chemistry

◆ Suggestions for Motivation or Enrichment:

1. Make a list of as many chemical elements as you can from memory and try to give one significant fact about each element you have named. Remember to include gases, liquids, and solids but do not include those substances that are a combination of more than one element.

2. Have someone describe an element without giving its name and see if you can identify the element he has in mind.

3. State the problems that an illustrator may have when he tries to picture chemical matters. Think in terms of relative sizes involved in some concepts and also of explanations involving reactions.

4. Presume that you are in charge of a farm and that you have the services of a good chemist available to you. Describe what you would hope to learn from him to optimize plant and animal reproduction and growth as well as to improve the farmer's personal health and welfare.

◆ Suggestions for Multimedia Resources:[*]

1. *God of the Atom*/VHS (Moody Videos, 28 min.). It explains modern atomic developments.

2. *Where the Waters Run*/DVD & VHS (Moody Videos, 28 min.). It explains how water sustains life.

3. *Roaring Waters*/VHS (Moody Videos, 28 min.). It explains how the power of water reshapes Earth's surface and the awesome forces of God's Creation.

◆ Suggestions for Supplementary Reading:[*]

1. Horn, C. J. "Water, Water, Everywhere" *Good Science*, October 1997. To read this article online, visit <http://www.icr.org/goodsci/bot-9710.htm>.

2. Wieland, Carl, and Jonathan Sarfati, Ph.D. "God and the Electron."*Creation* 21: 438–41. You may read the article online at <http://www.answersin genesis.org/creation/v21/i4/electron.asp>.

3. Fowler, Michael. 1997. "Evolution of the Atomic Concept and the Beginnings of Modern Chemistry." *Physics 252: Modern Physics* Web site. To read this article online, visit <http://galileo.phys.vir-ginia.edu/classes/252/atoms.html>.[**]

◆ Answers to Questions

Inside the Atom (Text page 28)
➤ Questions: Inside the Atom

1. The fundamental particles of an atom are *protons, neutrons,* and *electrons.* These particles differ from one another in their *mass* (weight) and *electrical charge.* Protons are positively charged particles, and electrons have an equal amount of negative electrical charge, so that the sum of the charges on one electron and one proton is zero. Neutrons have about the same mass as protons—about 2,000 times the mass of an electron—but have no charge.

2. The difference between an *isotope* and its corresponding element is the number of neutrons.

3. "Tracers" are radioactive atoms—isotopes that are unstable; this means the nucleus releases particles. These radioactive atoms are very helpful in biochemical reactions because the radiation (i.e., released particles) produced makes it possible to detect these atoms in far smaller amounts than in ordinary chemical tests.

➤ Taking It Further: Inside the Atom

1. *Answers will vary. The student is asked to choose an element from the periodic table (see page 33 of the textbook) not included in Figure 3-3 and list its number of protons, electrons, and neutrons.*

2. *Answers will vary. The student is asked to select one element from Figure 3-3 and research its chemical properties, its most common uses, and where it is most commonly found.*

3. It is possible to demonstrate that two particles have opposite electrical charges by allowing the two particles to move freely, as when suspended by threads; they will come together when they are held near each other because particles having opposite electrical charges attract each other.

Molecular Structure (Text page 30)
➤ Questions: Molecular Structure

1. The difference between an isotope and an ion is that an *isotope* is an atom of a particular element

---

[*] The listing of these suggestions does not necessarily imply endorsement of content.

[**] Professor Michael Fowler has other lectures at <http://gali-leo.phys.virginia.edu/classes/252/home.html>. He is a professor of physics at the University of Virginia.

that differs only by the number of neutrons, whereas an *ion* is an atom or group of atoms that has gained or lost one or more electrons. Atoms that have lost electrons are called *cations* (+). Atoms that have gained electrons are called *anions* (–).

2. When an ionic compound is dissolved in water (i.e., a process known as *dissociation*), the oppositely charged ions may separate. This process is known as *ionization.*

3. The two main types of bonds are *ionic* and *covalent.* An **ionic bond** is a chemical bond formed between two ions with opposite charges due to their mutual electrostatic attraction; it is an atom that is stabilized by gaining or losing an electron.

   A **covalent bond** consists of a pair of electrons, one furnished by each of the atoms involved in the bond, traveling in an orbit that includes the nuclei of both the atoms. It differs from ionic bonding in that electrons are shared between two atoms, but are not transferred outright to form ions. The pair of electrons orbiting around the nuclei of both atoms results in a strong chemical bond that is maintained intact in solution, since it does not depend on electrostatic attractions.

   Examples of each type of bond may vary. An example of an ionic bond is sodium chloride (NaCl, table salt). Examples of covalent bonds are oxygen (O_2), water (H_2O), and sugar ($C_6H_{12}O_6$, glucose).

4. Of the two types of bonds, the covalent bond is most commonly found in biological specimens; examples of covalent bonds are listed in #3 above.

5. The difference between a *substance* and a *mixture* is that a **substance** is some material that consists of only one kind of molecule (any portion of a substance will have the same physical and chemical properties as any other portion), but a **mixture** has varied properties throughout.

Teacher: For your information, *air* is a mixture of about 20% oxygen and 80% nitrogen, with traces of other gases. *Milk* is a complex mixture of water, protein, fat, and milk sugars. In contrast, *pure oxygen, water,* and *sugar* are examples of substances.

---

6. An *isomer* is a compound having the same molecular formula (the same kinds and proportions of atoms) but differing in the arrangement of those atoms. It is important to understand this concept because, with increasing numbers of atoms in a molecule, the number of possible isomers increases tremendously.

➤ **Taking It Further: Molecular Structure**

1. *Answers will vary. The student needs to find three Internet sites that pertain to molecular structure and summarize their content.*

2. *The student needs to draw the molecular structure of sugar (sucrose) using lines for bonds. See the model below or page 35 in the textbook.**

Inorganic and Organic Chemistry (Text page 34)

➤ **Questions: Inorganic and Organic Chemistry**

1. The unique properties of water are as follows:
   - Water is the only solvent that has the ability to readily dissolve materials with such widely different properties as inorganic salts and organic compounds.
   - Water has a higher heat capacity than any other common liquid, which means that a given amount of heat produces a smaller temperature rise in water than in other solvents.
   - Water molecules are cohesive, tending to cling together—a characteristic that most other solvents lack. Water also has the highest surface tension of any common liquid.
   - Water has the distinct property of being less dense in the solid (frozen) form than in the liquid. This means water molecules fit closer together in liquid form than they do in solid form, a most unusual situation.
   - Water also has the distinct ability to serve chemically as both an acid and a base.

   They are so important for sustaining life because:
   - The excellent solvent properties of water are particularly important because almost all chemical reactions in a living organism are carried out in a watery or aqueous environment, and not in a solid form.
   - Water's higher heat capacity helps to shield organisms against rapid and dangerous fluctuations in temperature, which also accounts for the moderation of climate near large bodies of water.
   - Surface tension causes water to rise to rather high levels when confined to a tiny capillary tube; this at least partially accounts for the ability of tall trees to draw moisture hundreds of feet into the air.
   - Due to the lower density of ice, organisms living in a lake are protected from freezing by the formation of a layer of ice above them.
   - Water's ability to serve chemically as both an acid and a base helps chemical reactions in living cells.

2. The difference between *inorganic* and *organic chemistry* is that organic chemistry deals with compounds that contain carbon atoms, and inorganic chemistry deals with all other compounds.

---

* Remind your student that carbon makes four bonds, oxygen makes two, and hydrogen only makes one. He may use the Internet or a chemistry textbook if necessary.

3. *Answers may vary. Only four compounds are required for each.* Organic compounds include proteins, nucleic acids, lipids, and carbohydrates; inorganic compounds include water, oxygen, nitrogen, and chloride and phosphate salts.

➤ **Taking It Further: Inorganic and Organic Chemistry**

1. *Answers may vary. The student is required to research Frederich Wohler and his contributions to science. In a short essay he must summarize Wohler's accomplishments and the importance of them.*

2. *Answers may vary. Since carbon's versatility provides the framework for the diversity found in nature, the student is asked to look up and draw several examples of the different ways carbon can form bonds.*

➤ **Questions: Chapter Review**

1. A solvent has the ability to dissolve other substances. When a solvent has a dissolved substance—or solute—in it, it is clear. The solute will go through filter paper and will not settle to the bottom. Substances that dissolve in water are *inorganic salts* like sodium chloride and *organic compounds* like sugars.

2. If one lived in a hot country away from civilization, it would not seem reasonable that water changes to a solid. An occurrence would seem reasonable if it was a common, repeatable phenomenon; but it is not necessarily so.

3. In an element (or substance) all the atoms are alike. The number of protons present in the nucleus of an atom defines which element it is; all atoms that have the same number of protons are considered to be the same element. However, variety can be found within an element due to varied numbers of electrons and neutrons.

4. Here the word *substance* is used in the ordinary, not the chemical sense. New properties may be added to the compound that the elements comprising it did not display.

5. Molecules that consist of only one kind of atom are known as *polyatomic elements* (e.g., O_2, or oxygen, which is a diatomic element), and a molecule that consists of more than one kind of atom is called a *compound* (e.g., H_2O, or water).

6. When a needle floats on water, the law of gravity is not broken. At the surface of water, the molecules are close together, making—by their cohesion—a kind of film. This cohesion is stronger than the gravity acting on the needle, making the surface film hold together. In this instance, gravity is not permitted to accomplish any action. Laws are not broken but give place, in certain situations, to other laws that are stronger.

7. *Chlorine atom* (Cl_2) is pale green gas; being a powerful oxidant, it is used in bleaching and disinfectants. It combines readily with nearly all other elements. As the *chlorine ion* (Cl^-) it is also the most abundant dissolved element in ocean water.

CHAPTER 4
Chemical Structure of Biological Materials
Text pages 35–39

Carbohydrates and Lipids
Proteins and Nucleic Acids

◆ **Suggestions for Motivation or Enrichment:**

1. Find or make a code for secret communication. Display a message written in the code and the key for solution. Discuss with other students the need for decoding properly, as for highway signs, dressmaking patterns, and medical prescriptions.

2. Learn from the appropriate office personnel just what is involved in computer printouts, especially for avoiding confusion or error. Collect samples that may be displayed.

3. Keep a record of your daily food intake, separating the items as chiefly carbohydrate, protein, or fat. Compare records with other students and make some group recommendations for general improvement.

4. Identify several "sugars" from unabridged general and medical dictionaries. Write down the names and formulae, and collect samples for tasting.

◆ **Suggestions for Multimedia Resources:***

1. *Biochemistry I.* Fogware Publishing. Visit their Web site at <http://www.fogwarepublishing.com>. CD-ROM; Windows OS only; common types of biological molecules; basic principles of cellular function; biological significance of carbohydrates and lipids; structure and role of glucose; the structural combinations of sugar molecules, glucose breakdown, and the role of ribose in photosynthesis and nucleoside triphosphate chemistry

2. Schmidel & Wojcik. 2004-2005. "Biochemistry Guide." *BioChem Hub.* To access these resources online, visit <http://biochemhub.com/biochem/chemistry.cfm>. This Web page provides an extensive list of links to Web sites having to do with biochemistry.

◆ **Suggestions for Supplementary Reading:***

1. Anson, Tom. 2004. "The Chemistry of Essential Oils ." *Anson Aromatic Essentials.* To read this article online, visit <http://www.therapeutic-grade.com/info/chemistry.html>.

---

* The listing of these suggestions does not necessarily imply endorsement of content.

2. Brown, Walt. "Chemical Elements of Life." To read this article online, visit <http://www.creation-science.com/onlinebook/LifeSciences33.html>. This is taken from Dr. Brown's book, *In the Beginning: Compelling Evidence for Creation and the Flood*, which refutes many evolutionary concepts.

3. Hurd, Gary S. 2004. "Ancient Molecules and Modern Myths." *The Talk.Orgins Archive.* To read this article online, visit <http://www.talkorigins.org/faqs/dinosaur/osteocalcin.html>. This Web site presents many challenging articles that will help the student understand the stakes involved in the creation-evolution controversy.[*] Note that the abbreviation YEC stands for Young Earth Creationists.

4. Malmos, Keith. 2004. "Carbon and the Molecular Diversity of Life." To read these course notes online, visit <http://faculty.valencia.cc.fl.us/kmalmos/hydrocarbons.htm>.[**]

◆ Answers to Questions

Carbohydrates and Lipids (Text page 37)

➤ **Questions: Carbohydrates and Lipids**

1. A *monomer* is the smallest unit that can still be identified as a specific organic compound; monomers can be chemically bonded into long chains or more complicated structures.

2. Both *starch* (complex carbohydrate) and *cellulose* (polysaccharide) have the same structural formula $(C_6H_{10}O_5)_x$[***]; they differ only in the manner in which the glucose units are attached to each other, *starch* having what is called α-glucosidic linkage and *cellulose* having the β-glucosidic linkage.

For extra credit:

Cellulose is the strong insoluble fiber found in the cell wall of plants (rarely in animals). On the other hand, starch serves as a storage pool of glucose, which can be rapidly degraded by a digestive enzyme to provide glucose when needed. Being completely resistant to this digestive enzyme, cellulose passes through the intestinal tract of animals undigested, with the exception of *ruminant* (chewing the cud) mammals, which have ruminal stomachs that contain cellulose-digesting bacteria that break down cellulose into usable sugars.

3. A common animal fat is tristearin (the crystallizable triglyceride $C_{57}H_{110}O_6$ of stearic acid).

4. Fat is *saturated* (having the greatest number of hydrogen atoms possible attached to the carbons—two per carbon) when the chain of carbons in the fatty acid chains are joined to each other with single bonds. Fat is *unsaturated* when one or more (polyunsaturated) double bonds form between the carbons in the chain. These double bonds are more easily broken down than their single bond counterparts.

➤ **Taking It Further: Carbohydrates and Lipids**

1. *Answers will vary. The student is required to build a molecular model of glucose using everyday items.*

2. *Phospholipids* closely resemble tristearin (animal fat) structurally, but in addition to the elements carbon, hydrogen, and oxygen, also contain **nitrogen** and **phosphorus**, making them more soluble in water than tristearin.

   Phospholipids perform an important role in biological membranes, due to the structure of **duel polarity**—a *polar head* (soluble in water) and a *nonpolar tail* (not soluble in water). Molecules of various types of fat tend to associate together because of their low solubility in water, but they are not found in the form of large covalently bonded polymers.

Proteins and Nucleic Acids (Text page 39)

➤ **Questions: Proteins and Nucleic Acids**

1. Proteins are so versatile because, when completely degraded, they yield about twenty different monomers—naturally occurring amino acids. These amino acids have widely differing chemical and physical properties, and these differences may be amplified when they are built into proteins.

2. The basic building blocks of living matter are *amino acids* (monomers of protein structure) and *nucleotides* (monomer units of nucleic acids); there is a minimum content of amino acids and nucleotides every cell must have to show the properties of life.

3. *Translation* is the process by which proteins are faithfully produced with identical structures over and over again in the life of an organism. The translation of the DNA language (based on only *four* kinds of letters) into the protein language (based on *twenty* letters) is the function of the other type of nucleic acid, RNA. Through a process called *transcription*, DNA is changed into RNA, which recognizes a certain sequence of three nucleotides as specifying a particular amino acid.

4. The similarity between proteins, nucleic acids, and carbohydrates is that they all are biopolymers that contain carbon, hydrogen, and oxygen atoms.

---

[*] As talkorigins.org home page states, "The primary reason for this archive's existence is to provide mainstream scientific responses to the many frequently asked questions (*FAQs*) that appear in the talk.origins newsgroup and the frequently rebutted assertions of those advocating intelligent design or other creationist pseudosciences." You will note the derogatory tone of their Web site, but we as Christians should not emulate this approach; the plain truth, presented in humility, is best.

[**] Professor Malmos's notes are from his course "Fundamentals of Biology I," based on chapter four of *Biology* (sixth edition) by Campbell and Reece. He teaches at Valencia Community College (<http://www.valencia.cc.fl.us/>).

[***] The variable *x* at the end of the formula indicates that this number varies, depending on the substance.

When completely degraded, these biopolymers yield various kinds of monomers.*

➤ **Taking it Further: Proteins and Nucleic Acids**

1. *The student is required to make a list of the twenty main amino acids found in most organisms and show their chemical structure. The following chart gives only seven out of the twenty. The* R *group indicates the variable structures of this part of the molecule.*

Amino Acid	Symbol 3 Lett.	Symbol 1 Lett.	R group
Aspartate	Asp	D	
Glutamate	Glu	E	
Lysine	Lys	K	
Arginine	Arg	R	
Histidine	His	H	
Tyrosine	Tyr	Y	
Tryptophan	Trp	W	

2. There are 4^3 or 64 three-letter combinations constructed from an alphabet of the four nucleotide letters—*T, C, A,* and *G.*

➤ **Questions: Chapter Review**

1. The suffix *-hydrate* comes from the Greek word meaning "water." However, the carbohydrates are not well named because the formula could be construed as six carbon atoms with a molecule of water attached to each (i.e., a *hydrate* of carbon, or carbohydrate). Actually, there are no water molecules as such in the structure, and the carbon atoms are joined to each other in a single unbranched chain.

2. *Isomers* are compounds with the same empirical formula but different structural formulas and properties. *Polymers* are chemical compounds or a mixture of compounds joined together through a

chemical reaction and consist essentially of repeating structural units.

3. Tristearin ($C_{57}H_{110}O_6$) has a large portion of carbon and hydrogen combined with a small portion of oxygen. Carbon and hydrogen combining with oxygen of the air give off much heat. Cellulose ($C_6H_{10}O_5$)$_x$ has small portions of carbon and hydrogen combined with a large portion of oxygen; O combining with O gives off no heat.

4. Experiments establish the essential facts; for instance, that a leaf gives off moisture. Chemical analysis is helpful, however. For instance, if a leaf in sunshine contains more starch than one in darkness, we conclude that light is needed in the process of making starch.

5. If all the food in an animal's stomach would pass into the bloodstream, diffusion alone would equalize the food in the intestine and the bloodstream. Evidently, another process is required.

CHAPTER 5

Chemical Transformations of Biological Materials
Text pages 41-44

Metabolism

Enzymes and Energy

◆ **Suggestions for Motivation or Enrichment:**

1. Explain what you mean when you say that you do not have enough energy to do a certain thing. Clarify your statements as related to a mountain climb and also to the writing of a term paper.

2. Research and be ready to describe the significance of a basal metabolism test, such as doctors use in diagnosis. If possible, inquire of someone who has experienced such a test to personalize the procedure and findings.

3. Analyze some advertising which emphasizes detergents using enzymes and report on the implications of such advertising.

4. Consult an unabridged dictionary for words beginning with *meta-, ana-,* and *cata-.* Write out the meanings of these prefixes and also of the words having biological interest.

◆ **Suggestions for Multimedia Resources:****

1. *Biochemistry II.* Fogware Publishing. Visit their Web site at <http://www.fogwarepublishing.com>. CD-ROM; Windows OS only; common types of biological molecules; basic principles of cellular function; biological significance of carbohydrates

---

* *Carbohydrates* yield only one kind of monomer (glucose), *nucleic acids* yield four kinds of monomers (nucleotides), and *proteins* yield about twenty kinds of monomers (amino acids).

** The listing of these suggestions does not necessarily imply endorsement of content.

and lipids; structure and role of glucose; the structural combinations of sugar molecules, glucose breakdown, and the role of ribose in photosynthesis and nucleoside triphosphate chemistry

2. Schmidel & Wojcik. 2004-2005. "Biochemistry Guide." *BioChem Hub.* To access these resources online, visit <http://biochemhub.com/biochem/chemistry.cfm>. This Web page provides an extensive list of links to Web sites having to do with biochemistry.

◆ Suggestions for Supplementary Reading:*

1. *Energy Information Portal.* Department of Energy: Energy Efficiency and Renewable Energy (EERE). Visit <http://www.eere.energy.gov/>. This site is a gateway to hundreds of Web sites and thousands of online documents on energy efficiency and renewable energy. In addition, the EERE sponsors "Student Vehicle Competitions"; for more information, visit <http://www.eere.energy.gov/afdc/resources/kids_competitions.html>.

2. "Diagnostic Tools to Measure Building Performance." *The Energy Conservatory* (*TEC*). Visit *TEC*'s site at <http://www.energyconservatory.com/articles/articles1.htm>. Articles, newsletters, and linked articles on energy conservation

3. "Symposium: Catalysis." *ScienceWeek*, June 13, 2003, Vol. 7 Number 24A. To read this article online, visit <http://scienceweek.com/2003/sw030613.htm>. This article provides etymological and historical information on catalysis.

◆ Answers to Questions

Metabolism (Text page 41)

➤ **Questions: Metabolism**

1. *Metabolism* is all the organism's chemical reactions combined—the chemical changes in living cells by which energy is provided for vital processes and activities and new material is assimilated.

2. *Anabolism* is all the organism's chemical reactions leading to the buildup of cellular components; *catabolism* is all its degradative reactions.

3. The same types of reactions, which occur inside an organism, can occur outside of it. For example, the oxidation of fat outside a cell occurs only at a high temperature (fat dripping from meat onto coals). The same reaction occurs quite readily at body temperature in a living organism. The difference is that the organism is able to utilize some of the energy produced in the reaction to accomplish useful work, such as muscular contraction, synthesis of proteins, or transmission of nerve impulses.

➤ **Taking It Further: Metabolism**

1. *The student should make a list of anabolic and catabolic reactions that occur in the body and list the products formed. The list should include **catabolic reactions** that oxidize (or partially oxidize) and degrade fuel molecules from fats, carbohydrates, and proteins to obtain energy (fatty acids, glucose, and amino acids) for driving the various processes of life and **anabolic reactions** that synthesize the essential molecules of the cell (DNA, RNA, protein, lipids, glycogen, etc.).*

Enzymes and Energy (Text pages 43–44)

➤ **Questions: Enzymes and Energy**

1. The main function of enzymes is to catalyze specific biochemical reactions at body temperatures. They are so important because without a catalyst a particular chemical reaction would proceed slowly.

2. The main characteristics of an enzymic catalyst are as follows:

   • Enzymes are very *sensitive* to their environmental conditions. Moderate increases in temperature or changes in acidity may result in complete and irreversible loss of its ability to act as a catalyst.
   • Enzymes are marvelously *efficient*. Their high efficiency makes it possible for only a very small amount of enzyme to meet the needs of the organisms.
   • Enzymes are quite *specific* with respect to the type of reaction they catalyze. There is a different enzyme for every reaction that occurs in a cell.

3. ATP is a naturally occurring, high-energy compound called *adenosine triphosphate.* It is important because, in cellular respiration, much of the chemical energy involved is stored as either ATP or as compounds readily converted (enzymatically) to ATP, to be utilized when needed by the organism.

4. The equation of cellular respiration and the equation of photosynthesis may well be called the "equations of life." Plants are able to store energy from sunlight as chemical energy in compounds, such as **glucose**, by the process of *photosynthesis*. Likewise, man and animals can store energy from foodstuff as chemical energy in compounds, such as **ATP**, by the process of *cellular respiration*.

➤ **Taking It Further: Enzymes and Energy**

1. *The student is required to make a list of three organic and inorganic catalysts and name the reactions that they promote. The catalysts should include the following categories:*

   • vitamins, enzymes (*organic*); some minerals (*inorganic*)
   • phase-transfer catalysts, oxidation catalysts, acidic catalysts for protection and deprotection reactions
   • acidic catalysts for C–C bond-forming reactions
   • Metabolic enzymes that run the body are synthesized from raw food (fruit, vegetables, nuts, grain, and even raw meat). All vitamins are enzymes.

---

* The listing of these suggestions does not necessarily imply endorsement of content.

➢ **Questions: Review**

1. Potassium chlorate, when heated, gives off oxygen slowly; mixed with manganese dioxide, oxygen is given off more rapidly. Manganese dioxide is a catalyst. Such a substance speeds up a chemical reaction without reacting permanently with any of the substances involved. After the reaction, the catalyst is found to be unchanged.

2. *Answers may vary. Only four forms of energy are required.* Energy may come in the form of heat, light, motion, position, chemical energy, or electricity.

3. A sand bar in a stream sometimes gains sand, sometimes loses it. Gain and loss of sand are due to movement of water without change in the sand except in position. This is physical change, whereas metabolism consists of chemical changes in the structure of the substances.

4. Indirectly, but in a very real sense, the lion depends upon plants. It eats animals, such as the antelope, whose food is plants.

5. Respiration, like burning, is a union of oxygen with some substance, and both processes give off heat. The name, "wet burning," indicates that respiration is not stopped by water. But the student should learn early that respiration depends upon life and is very complex.

◆ **Think-Session Guide for Unit 2**

Chemical Perspectives in Biology

> *Subject: Analysis of serial causes*
>
> *Purpose: To see the broad picture of chemistry-based biology*

Teacher:	This "search for order" is designed to help the student overcome a fear of chemistry. He should, at the end of the unit, realize that a knowledge of the chemistry of a biological process is basic to an understanding of its biology.

a.	**To the student:** It has been said that the most important element in reducing traffic accidents is the "nut behind the wheel." Analyze the accident-nut sequence for smaller sequences in between.

Teacher:	Traffic accidents involve cars. Cars are designed, built, driven, and maintained by people. The driver of the car is responsible for the "behavior" of his vehicle.

b.	**To the student:** Do the same thing for the series: "A hamburger gives its energy."

Teacher:	The meat and bread are physically broken down, then chemically broken down (digested). The small molecules are then redistributed, stored, or "burned up" to release energy. Obviously, this can be expanded to the limit of detail necessary to introduce the chemistry of Unit 2.

UNIT 3
The Continuity of Life

CHAPTER 6
The Nature of Living Things
Text pages 47-62

More Than Chemistry

Cells

Law of Biogenesis

◆ **Suggestions for Motivation or Enrichment:**[*]

1. Describe a microscope, pointing out the main parts and tracing the path of light and the production of a magnified image.

2. Consult an unabridged dictionary for words beginning with *proto-*, and write out the meaning of the prefix and of the words. Underline those having biological implication.

3. Describe how you would try to prove to someone that life does not originate from some kind of soup left in a jar.

4. Recall from ancient mythology some of the startling beliefs which societies have held regarding the origin of people on earth. Acquaint yourself with one or more stories of origin still honored by some Native American tribes. Make a list of names for the supreme power or person, apart from Judeo-Christian literature. Compare these myths with the factual account of Genesis.[**]

◆ **Suggestions for Multimedia Resources:**[*]

1. *Cell Biology I.* Fogware Publishing. Visit their Web site at <http://www.fogwarepublishing.com>. CD-ROM; Windows OS only; prokaryotic and eukaryotic cells, structure and function of organelles in plant cells, phases of the cell cycle, the states of mitosis, cell division

2. *Cell Biology II.* Fogware Publishing. Visit their Web site at <http://www.fogwarepublishing.com>. CD-ROM; Windows OS only; how lipid and protein molecules assemble to form cellular membranes, fluid mosaic model and other major pathways for transporting molecules through membranes, systems of cell motility based on

---

[*] The listing of these suggestions does not necessarily imply endorsement of content.

[**] Jerry Bergman's article, "The Origin of Creation Myths," provides excellent background information for this exercise. Go to <http://www.creationism.org/csshs/v06n2p10.htm>.

Also, the "Creation/Migration/Origin Stories" Web page may be of some help concerning certain Native American myths. Go to <www.indians.org/welker/legend.htm> for more info.

---

microtubules and microfilaments, role of microtubule sliding to produce flagella motion and chromosome movement

3. *DNA: The Instruction Manual for All Life.* TheTech. Visit their Web site at <http://www.thetech.org/hyper/genome/>. This Web site provides basic information about DNA and the human body.

4. *The Cell Nucleus.* University of Texas Medical Branch. Visit their Web site at <http://cell-bio.utmb.edu/cellbio/nucleus.htm>. This site provides information about the structure and function of the cell nucleus.

5. Balkwill, Dr. Fran, and Mic Rolph. *Cells Are Us.* Cancer Research UK. Visit their Web site at <http://www.icnet.uk/kids/cellsrus/cellsrus.html>. This site uses animation to illustrate cell division and human growth.

6. *The Virtual Cell Web Page.* Visit their Web site at <http://www.ibiblio.org/virtualcell/>. This Web site provides a wealth of information about cell biology, using both text and graphics.

7. Dalton, Mark. *The WWW Cell Biology Course.* Visit their Web site at <http://www.cbc.umn.edu/~mwd/cell.html>. An online biology book, with an emphasis on cell biology; still a work in progress

◆ **Suggestions for Supplementary Reading:**[*]

1. Bergman, Dr. Jerry. "Why Abiogenesis Is Impossible." *Creation Research Society Quarterly*, 36: 4. To read this article online, visit <http://www.creation-research.org/crsq/articles/36/36_4/abiogenesis.html>. Critiques the idea that life can arise spontaneously from non-life molecules under proper conditions

◆ **Answers to Questions**

More than Chemistry (Text page 50)

➤ **Questions: More than Chemistry**

1. In 1828, the first biological substance produced through chemical means in a laboratory was urea. Frederich Wohler was the scientist that accomplished this groundbreaking feat.

2. Joseph Priestley's work with **mercuric oxide** (HgO) was so important because he found that a mouse lived longer in this "new" gas (*oxygen*, produced by heating HgO) than it could have lived in the same volume of ordinary air. When he learned that this gas did not harm animals, he breathed some of it himself and found it very refreshing. In this gas, a candle burned rapidly with a bright, noisy flame.

3. An understanding of chemistry is foundational to an understanding of biology because all of life is dependent on the chemical reactions that take place within all living things. Creatures and plants are composed of matter, and their environment influences them.

➤ **Taking it Further: More than Chemistry**

1. *Answers may vary.* Life is more than chemistry because, even though creatures and plants are composed of matter, in their being there seems to be something more. They are under their own unique control. *The following examples are given in the text, but the student only needs to give two or three:*

   • production of digestive enzymes (which is affected by emotions)
   • gastric juice readily digests tough bits of meat but does not digest the stomach wall
   • plants and creatures take in inorganic material and use it in the organic functions of their cells
   • matter comes and goes, but the individual form of a creature or plant continues
   • some actions seem to be initiated by creatures' own desires, and it is difficult to classify them as results of stimulation by outside forces
   • reproduction seems to be beyond the province of physics and chemistry
   • placement of hormones at the right place at the right time cannot be explained chemically
   • though chemistry may explain the growth of creatures and plants, they cease to grow at a certain size

2. *Answers will vary. The student is asked to look in current periodicals for articles that discuss the chemical explanation for emotions affecting reactions in the body and write a summary of his findings.*

Cells *(Text page 58)*

➤ **Questions: Cells**

1. The *cell theory* is the concept that cells are the fundamental structures of which organisms are built and through which they function. All living organisms are made up of one or more cells, which carry on the processes and activities of life; and cells only come from preexisting cells. This theory is important in that it accurately explains that diseased cells arise from preexisting diseased cells, not from an imbalance of the four "humors," or bodily fluids.

   For extra credit: Your student may list and describe the major parts of a cell.

   • **Boundary:** thin, protective membrane that surrounds the protoplasm; provides shape, protection, cellular communication, and transport
   • **Protoplasm:** living matter in the cell, both the nucleus and cytoplasm
   • **Cytoplasm:** portion of the protoplasm that lies outside the nucleus; supported by a matrix of microfilaments; holds the majority of the cellular organelles; location for most metabolic activities of the cell

   • **Nucleus:** the spherical body that contains the chromosomes; control center of the cell

2. The seven structures within a cell are as follows:

   • **Endoplasmic reticulum:** a series of membranes that form canal-like sacs, found throughout the cell; conducts materials through the cell in an orderly way and provides a working surface for enzyme action; also provides structure to the cell
   • **Ribosomes:** circular strands of messenger RNA that are involved in forming proteins
   • **Mitochondria:** powerhouses of the cell; slipper-shaped organelles that release energy in the form of ATP from food; surrounded by duel membrane; where cellular respiration takes place
   • **Golgi body:** consists of flattened membrane sacs; "shipping department" of the cell; synthesizes and puts together macromolecules (biopolymers); stores certain biopolymers
   • **Lysosomes:** saclike structures, surrounded by a single membrane; full of digestive enzymes and are used by the cell to break down food and destroy harmful bacteria or viruses entering the cell; rupture and dissolve a cell so that a new cell can take its place
   • **Vacuoles:** contain various nonliving materials; some help to regulate the amount of water in a cell; prevents plasmolysis*; engulfs and usually destroys particulate matter; eliminates waste, provides storage, and transports materials through the cell
   • **Centrioles:** small bodies composed of microtubules (typically nine groups of three) and located near the nucleus; involved in cell division

3. Differences in plant and animal cells are found in the *boundary* and the *cytoplasm*. The boundary of a plant cell has a prominent outer wall composed of cellulose and a thin inner membrane; animal cells only have a thin membrane for a boundary.

   The cytoplasm of a plant cell usually surrounds a central cavity, or vacuole, containing *cell sap*. The cytoplasm has two types of plastids—*leucoplasts* (colorless plastids that store energy in the form of starch and other nutrients) and *chromoplasts* (pigmented cells that carry out photosynthesis).

4. The names of the phases of mitosis are the *interphase* ("in-between phase"), *prophase* (centriole divides and aster fibers lengthen), *metaphase* ("transformation phase"), *anaphase* (cytokinesis), and *telophase* ("last phase").

5. *Mitosis* is division of chromosomes preceding the division of cytoplasm; *meiosis*, however, is the type of cell division in which there is reduction of chromosomes to the monoploid number during oogenesis (formation and maturation of the egg) and spermatogenesis (process of male gamete formation).

6. Forty-six is the *diploid number* for man, but a human sperm has just twenty-three chromosomes, the *monoploid number.*

---

* *Plasmolysis* is the shrinking of the cytoplasm away from the wall of a living cell due to outward osmotic flow of water.

➤ **Taking it Further: Cells**

1. *The student is asked to draw and describe each phase of mitosis and meiosis, making sure to label all of the important structures. See the diagrams below.*

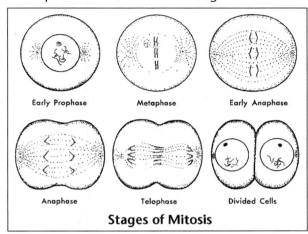

Early Prophase   Metaphase   Early Anaphase

Anaphase   Telophase   Divided Cells

Stages of Mitosis

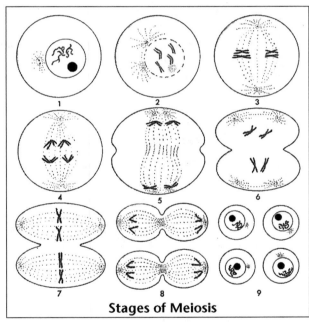

1   2   3

4   5   6

7   8   9

Stages of Meiosis

2. Ribosomes are formed in the nucleus and transported through the cell by the endoplasmic reticulum. Rough endoplasmic reticulum gets its name from ribosomes attached to the membrane; ribosomes can also be found free in the cytoplasm. They are circular strands of messenger RNA that are involved in forming proteins.

Law of Biogenesis (Text page 62)

➤ **Questions: Law of Biogenesis**

1. The difference between a *law* and a *theory* is that a **theory** is a group of general propositions offered to explain a broad range of phenomena for which an abundance of evidence has been accumulated; a **law**, however, is primarily a well-established *hypothesis* that has been extensively tested.

2. The *Law of Biogenesis* is the biological principle that life arises from life, but the idea of *spontaneous generation* claims that organisms developed from lifeless materials. If one relies on presently known processes, there is no doubt that life is generated only by reproduction from life. Historically, biogenesis has been so well established through the facts revealed by experimentation that it is recognized as a law; moreover, it has discredited the concept of spontaneous generation of life.

3. Confirmation of scientific findings from more than one source is important before any experiment is considered valid. If experiments can be repeated with the same results, then an agreement can be reached among scientists that the given hypothesis is valid.

4. The main contributors to the development of the Law of Biogenesis are **Francesco Redi**, who demonstrated that flies are not generated spontaneously in meat but come from other flies; **Louis Joblot**, who concluded that material freed of all life could not generate living things; **John Needham**, who showed that living things did appear in closed as well as open vessels, but **Lazzaro Spallanzani** showed that Needham's experiments were flawed; **Nicholas Francois Appert**, who developed the method of canning fruits and vegetables by heating them and sealing them in cans to keep out the air; **Lord Joseph Lister**, who devised a can of dilute carbolic acid to make a fine spray over the patient and surgeons during an operation to reduce the number of infections; and **Louis Pasteur**, who performed a number of well-planned experiments which proved that life could not develop of itself, even in the presence of air.

➤ **Taking it Further: Law of Biogenesis**

1. *Answers may vary. The student is asked to outline Redi's and Pasteur's experiments with flies that showed spontaneous generation did not take place.*

Teacher: The sections in the text that cover this material are *6-19 Belief in Life from Life* (pages 58–59) and *6-23 Conclusive Demonstration* (pages 60–61).

Redi's experiments:

(a) Life history of the fly

(b) Meat and air and flies

(c) Meat—air—flies

(d) Meat and air—flies

Pasteur's experiments:

(a) Bacteria excluded by curved tube

(b) by cotton

(c) bacteria scarce in high altitude

(d) bacteria live in air

2. *Answers may vary.* The Law of Biogenesis shows that life does not spontaneously arise in its present forms from non-life in nature. No life has ever been observed to arise from non-living matter, thus evolutionists are unable to answer how life did arise in the past, based on their own theory of uniformitarianism.[*] And yet, they still advocate that some kind of *spontaneous generation* or *abiogenesis* ("non-biological origins") occurred billions of years ago.

➤ **Questions: Chapter Review**

1. People thought there was a mysterious power in life which digested food, and that this power had to work within the human body. Lazzaro Spallanzani showed the power of life is actually manifested in the *production* of gastric juice.

2. *First,* growth of an animal is not limited to the outside. *Second,* it involves a chemical change in the substance which is added. *Third,* the added material becomes organized into cells and tissues.

3. Chemical analysis kills living flesh, and at death the organization changes quickly. Therefore, its actual structure while alive is, to an extent, an inference.

4. A plant ceases to grow at death, when environmental conditions are unfavorable, and when any part reaches a size which we call maturity.

5. We know that fire is essentially union rather than destruction because its products, carbon dioxide, water, and ashes can be recovered and tested.

6. Chemical action can be predicted because it is always the same under similar circumstances. On the other hand, animals have a choice of reactions.

7. *Mitosis* is essentially an orderly division of a cell into two daughter cells. Of course, this process involves movement, but not for its own sake.

8. Hooke discovered that cells exist in some plant material; Schwann added that cells are found in all living tissue and are the basic units of living structure.

9. A branch of some plants may be removed and, when placed in soil, will add roots, thus becoming a new plant. Some buds fall off and grow into plants; e.g., the tiger lily. Some plants send tip sprouts from the roots, and these sprouts then grow roots of their own and become mature plants. Small bulbs form from large ones.

10. Plants do not have a conscious purpose of forming food. They do, however, form food for plant-eating animals, and these animals may be eaten by others.

11. Mitosis, along with other necessary processes, at present is well-fitted to the needs of living things. Intermediate processes would not be well-fitted and would cause the death of plants and animals.

12. A common belief in the early nineteenth century was that decay and fermentation, and even some infections, occur naturally. Then, when decay starts, bacteria are attracted, since this gives them a favorable place to live. People were familiar with decay and fermentation, but they could not see the tiny microorganisms in the air.

13. Scientists have demonstrated that logic alone does not always inform one about the actual world. Observation is quite necessary. Specifically, since it seems illogical that maggots change into flies, Redi could not depend upon reason alone and resorted to observation to verify his conclusions.

14. There was no man on earth to observe its formation; and, if experiments are devised to test the creation, we cannot be certain that they test conditions as they were at that time. This indicates the worth of a revealed record.

15. This chapter has taught that, where careful scientific methods can be used, scientific observation and experimentation are valuable. They are useless if they are not carefully performed.

CHAPTER 7
The Science of Genetics
Text pages 65-91

Heredity
Applied Genetics
The Chemical Basis of Gene Action
Population Genetics
Genetics and the Species

◆ **Suggestions for Motivation and Enrichment:**

1. Sketch a family tree of your relatives, according to their birth relationships, and indicate some strong characteristic according to its manifestation within a family group (e.g., shape of a facial feature or hair color).

2. List as many "freaks of nature" as you know about or have seen. Note which ones seem to have had inherited possibilities and prepare to defend your conclusion.

3. Make an inquiry of an appropriate agency or individual regarding the items which are considered important in placing a child for adoption with families that wish to maintain resemblances to natural

---

[*] *Uniformitarianism* claims that the same processes that shape the universe occurred in the past as they do now, and that the same laws of physics apply to all parts of the knowable universe; thus, present-day processes can be used to interpret past patterns, as evolutionists claim. By contrast, *catastrophism* states that the surface features of Earth originated suddenly in the past, by geological processes radically different to those currently occurring, as creationists claims.

members. Evaluate this concept and modern adoptive standards in an essay.

4. Research the process and importance of blood grouping for blood transfusions and for family planning. If possible, discover your own blood group or that of someone known to you, and compare with other family members.[*]

◆ **Suggestions for Multimedia Resources:**[**]

1. *Talking Glossary of Genetic Terms.* National Genome Research Institute. Visit their Web site at <http://www.genome.gov/glossary.cfm>. It provides a searchable glossary of genetic terms.

2. *The Genetic Science Learning Center.* Eccles Institute of Human Genetics. Visit their Web site at <http://gslc.genetics.utah.edu/>. This is an online genetics curriculum.

◆ **Suggestions for Supplementary Reading:**[**]

1. Lester, Lane. "Genetics: No Friend of Evolution." *Creation*, 20: 220–22. To read this article online, visit <http://www.answersingenesis.org/creation/v20/i2/index.asp>. This article critiques evolution in light of modern genetics.

2. Johansen, Mark. "The Living Database: A Software Engineer Looks at God's Blueprint for a Human Being." *Creation*, 22: 433–35. To read this article online, visit <http://www.answersingenesis.org/creation/v22/i4/database.asp>.

3. Sarfati, Jonathan. "DNA: Marvelous Messages or Mostly Mess?" *Creation*, 25: 428–31. To read this article online, visit <http://www.answersingenesis.org/creation/v25/i2/dna.asp>.

◆ **Answers to Questions**

Heredity (Text page 70)

➤ **Questions: Heredity**

1. *Answers may vary.* The color of people's eyes is an example of a trait that is only influenced by genetics, and whether or not a person will develop diabetes is influenced by both environment and genetics.

2. *Answers may vary. The student is asked to describe Gregor Mendel's laws of heredity in his own words.* The first law of Mendel, the **law of segregation**, deals with genes of the same pair that are alleles; in the formation of gametes, genes that are alleles segregate—genes of the same pair go into different gametes, not into the same gamete. The second law of Mendel, the **law of independent assortment**, deals with genes of different pairs (not alleles); in the formation of gametes, genes that are not alleles assort at random.

3. *Incomplete dominance* is the blending of phenotypes, where the hybrid is some sort of blend between the homozygous parent types. *Codominance*, however, is the full expression of two phenotypes (one from each parent), where neither one is dominant or recessive to the other; the hybrid is a mixture of the heterozygous parent types.

4. If the parents were both heterozygous (*Tt*) for a trait, the probable phenotypic and genotypic ratio of children would be 3:1.

	T	t
T	TT	Tt
t	Tt	tt

5. Ways genetic material can be altered experimentally are by the use of high-energy radiation, heat, or certain chemicals. Mutations that occur in nature are most likely caused by radiation. The one way that is least likely to cause a mutation is heat.

➤ **Taking It Further: Heredity**

1. *Answers will vary. The student is asked to find a scientific paper on the study of genetics and write a summary of the conclusions. Internet sources may be used.*

2. The low probability of a helpful mutation does not support the modern theory of evolution. It is less likely that such a mutation would create a new or more complex species. Moreover, a mutation that produces a change considered advantageous in some unusual environment generally has injurious physiological effects that weaken the individual. Thus, even rare mutations that might be described as good for the species may actually be bad.

Applied Genetics (Text page 78)

➤ **Questions: Applied Genetics**

1. Geneticists are typically trying to develop three things in organisms: desirable quality for a specific purpose, disease resistance, and ability to grow in a specific environment.

2. A test cross or backcross is a cross of a hybrid with one of its parents, thus the geneticist then knows his organism is pure. This is important to the study of genetics because, by breeding only the pure individuals, the desired breed can be established.

3. *Answers may vary.* Examples of work in the field of genetics improving people's daily lives are many.

   One is the selective breeding of cattle. Cattle are bred for high milk production or high butterfat production; cattle are also bred for quality beef. Hornless cattle are desired because they do not injure each other or the men handling them. The Santa Gertrudis cow combines heat resistance of

---

[*] For more facts, read "Blood Groups, Blood Typing and Blood Transfusions" at <http://nobelprize.org/educational_games/medicine/landsteiner/readmore.html>.

[**] The listing of these suggestions does not necessarily imply endorsement of content.

the Brahman cattle and good meat qualities of the Shorthorn breed. A breed developed by crossing Brahman and Angus is called Brangus.

Genetics has also improved people's daily lives in the breeding of plants. Wheat has been developed that produces good flour, will ripen early to avoid being killed by early frost, and is resistant to rust. Cotton breeders removed the fuzz from cotton leaves so they would not hang in the mechanical cotton pickers. Commercial strawberries were produced by crossing tiny, soft berries with luscious flavor (from North America) with large, firm berries with inferior flavor (from the Andes Mountains of South America). *Burpee stringless green pod* was developed by bringing together such characteristics as thick, tender pods and the absence of strings.

4. A *sex-linked trait* is a trait whose genes are carried on the sex chromosomes.

5. Inbreeding increases the risk of inheriting disease because every generation after the F_2 will be the same as the F_2.

➤ **Taking It Further: Applied Genetics**

1. *The student is asked to develop his own dihybrid cross for eye color and height. The following chart shows an example of the results. (Letters used as abbreviations may vary.)*

	EH	Eh	eH	eh
EH	EEHH	EEHh	EeHH	EeHh
Eh	EEHh	EEhh	EeHh	Eehh
eH	EeHH	EeHh	eeHH	eeHh
eh	EeHh	Eehh	eeHh	eehh

2. *Answers will vary. In a short essay, the student is asked to discuss evolution as a belief system instead of a scientific theory.*

The Chemical Basis of Gene Action (Text page 86)

➤ **Questions: The Chemical Basis of Gene Action**

1. Purines and pyrimidines are complex ring-shaped structures made up of carbon atoms alternating with nitrogen atoms. Pyrimidines contain only one ring (four carbon and two nitrogen atoms); purines are double-ring structures—the additional ring has one carbon and two nitrogen atoms.

2. *Dehydration synthesis* is the process by which monomers of organic compounds join together to make polymers; e.g., component parts of DNA reassemble to form more complex molecules—adenine, guanine, cytosine, and thymine nucleotides.

3. Adenine ($C_5H_5N_5$), guanine ($C_4H_6N_5O$), cytosine ($C_4H_5N_3O$), and thymine ($C_5H_6N_2O_2$) are the nucleotides of DNA. Adenine ($C_5H_5N_5$), guanine

($C_4H_6N_5O$), cytosine ($C_4H_5N_3O$), and uracil ($C_4H_4N_2O_2$) are the nucleotides of RNA.

4. *Transcription* is the process in which a single strand of DNA builds on itself a complementary sequence of messenger-RNA; *translation* is the process in which messenger-RNA controls the sequence of amino acids in a protein by combining with complementary transfer-RNA units that carry amino acids.

5. Mutations have helped scientists better understand that the DNA mechanism is highly specific and integrated; any tinkering with DNA leads to nonsense codes and an inferior organism.

➤ **Taking It Further: The Chemical Basis of Gene Action**

1. *The student is asked to transcribe and translate the the DNA strand—TCAGAATACTCTATT—into mRNA and proteins. (See pages 83 and 84.)*

   • The transcription of this DNA strand to the mRNA is *AGUCUUAUGAGAUAA.*
   • The translation of the above mRNA into protein is *AGU* (serine), *CUU* (leucine), *AUG* (methionine), *AGA* (arginine), *UAA* (stop).

Teacher: Note that the *UAA* is a codon, or stop instruction, which is not specifically stated in the text.

---

2. *The student is asked to outline all the steps used in the above question. See page 84 of the text.*

Population Genetics (Text page 87)

➤ **Questions: Population Genetics**

1. If one person out of 50,000 is homozygous for a recessive trait, the frequency of this recessive trait is $1 \div 50,000$ or 0.00002. To find the frequency of the gene for this recessive trait in the population, one must determine the square root of 0.00002, which is approximately 0.004. This means that of every 1,000 genes in the population, four are genes producing the recessive trait and 996 are genes producing the normal trait. Therefore, 996 people out of a thousand will be homozygous dominant. (*See page 87 in the text.*)

2. With most traits, equilibrium is finally reached between factors reducing the frequency of a given gene and those that increase its frequency. From then on, the frequency of the genes and the phenotypes remains constant.[*] Mutation and migration, however, prevent equilibrium.[**]

---

[*] The Hardy-Weinberg Law states that, if mating is random and no external factors are involved, the gene frequencies and genotype proportions in a population remain constant from generation to generation.

[**] Factors that can disturb this equilibrium are (1) high radiation, which can produce mutations that are harmful or even lethal, and (2) the effective treatment of some genetic disorders, which allows heteromorphic individuals to pass their defective genes along.

➤ **Taking It Further: Population Genetics**

1. *Answers will vary. The student is asked to discuss the balance between medical advancements and genetic stability and the importance of this issue.*

Genetics and the Species (Text pages 90–91)

➤ **Questions: Genetics and the Species**

1. The purpose of eugenics is allegedly for improving the human species through genetic engineering by upgrading the genetic basis for intelligence and eliminating the undesirable genes.

2. Some of the problems in accomplishing this purpose are: intelligence is complex and has not been fully defined; it is not known just what is to be raised, nor can one be sure that what he measures in intelligence tests are the only factors in intelligence. Also, the genetic basis of intelligence is little understood. In addition, if it would be desirable to eliminate low intelligence, it is not a simple matter since most defective genes are recessive and mutations are difficult.

3. Moral questions are faced by those trying to implement eugenic principles:

   • since all individuals carry some recessive defective genes, judging which individuals are "more fit" to become parents is a serious moral issue
   • government control of eugenics also has its dangers, exemplified in Hitler's extermination of millions of Christians and Jews that he considered "unfit"
   • there are serious moral considerations in denying parenthood to anyone, since some very famous and useful people have been physically heteromorphic

➤ **Taking It Further: Genetics and the Species**

1. If the probability of being heterozygous for a certain trait is 1 in 100, the probability of a child showing the recessive trait is as follows:

$$\frac{1}{100} \times \frac{1}{100} \times \frac{1}{4} = \frac{1}{40,000}$$

2. *Answers will vary. The student is asked to research a genetic disorder and write a short essay on its cause and effects.*

➤ **Questions: Chapter Review**

1. Environment affects the phenotype. For instance, a plant growing in sunlight becomes deep green; a person becomes tanned. If fall-out from a nuclear explosion causes a mutation, this is a change in a genotype, but such a change is very rare.

2. The genotype of homozygous yellow peas is YY; of heterozygous yellow peas, Yy. The homozygous peas can transmit only one type of gene for color, namely Y. We place it at the left of our rectangle. Heterozygous peas can transmit two genes for color, namely Y and y (green). We place them at the top of our rectangle and combine them as genes would combine in fertilization of the flowers if they are crossed.

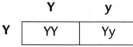

Genotype of offspring: one half *YY*, one half *Yy*. Phenotype of offspring: one half *yellow homozygous*, one half *yellow heterozygous*.

3. We assume the black parents are homozygous for color and we are sure the white parents are homozygous, else they would not be white. The first filial generation are all black. (*See answer 2.*) The second filial generation comes from mating among the first filial generation, Bb x Bb, and gives roughly three-fourths black and one-fourth white. These are phenotypes.

4. Green in pea seeds has been found, by test mating, to be recessive. If heterozygous, the other color present would dominate and be expressed.

5. Polled cattle, seedless oranges, stringless green beans, and beardless wheat are examples of mutations that are of benefit to man.

6. A well-informed biologist knows that accumulation of desirable genes, as in beets and corn, reaches a limit, after which there is no further change. The effect of selection is much too limited to account for the origin of mammals.

7. Inheritance of a habit would be an example of Lamarck's acquired character inheritance. Tests to prove such inheritance do not give positive results.

8. The expression "Like begets like" is so condensed that it actually says nothing. It is usually taken to mean that there is a tendency for children, mammals, and plants to resemble their parents.

9. The minutes of the meeting record that there was no discussion of Gregor Mendel's paper on peas, although now it is considered well-written indeed. The minutes record further that, later in the evening, a member mentioned with enthusiasm a book written six years previously by a certain Englishman, Charles Darwin, and this is what they discussed. It was the same all over Europe for 35 years. Apparently, the significance of Mendel's law was not understood.

10. Transcribe this DNA strand into RNA and translate the RNA into amino acids: *TCAGTACAGTACATT*.

    • The transcription of this DNA strand to the mRNA is *AGUCAUGUCAUGUAA*.
    • The translation of the above mRNA into protein is *AGU* (serine), *CAU* (histidine), *GUC* (valine), *AUG* (methionine), *UAA* (stop).

Teacher: Note that the *UAA* is a codon, or stop instruction, which is not specifically stated in the text.

11. The Central Dogma of molecular biology states that information from DNA can be used to specify synthesis of more DNA (the process of *replication*) or synthesis of RNA (the process of *transcription*). The information from RNA can then be used to specify synthesis of specific proteins (the process of *translation* of a four-letter nucleotide code into a twenty-letter amino acid code). According to the Dogma, information transfer is in only one direction; information from protein never specifies RNA or DNA sequences, and RNA never specifies DNA.

12. The study of population genetics is important to geneticists because of the relative frequency of different genes in the population and the changes that come about in the relative proportions of genes.

13. *Answers will vary. The student is asked to read George Grant's booklet,* Killer Angel: A Biography of Planned Parenthood's Margaret Sanger *and discuss the effect of eugenics on Margaret Sanger's life.*[*]

CHAPTER 8
The Development of the Individual
(Text pages 93-99)

Growth of the Embryo
Experimental Embryology

◆ **Suggestions for Motivation or Enrichment:**

1. Note the displayed items in a well-stocked meat market or butcher shop and list the distinct kinds of body structures that had to develop in order to produce complete animals.

2. Using a college dictionary, write out the meanings of these prefixes and some representative words: *ecto-*, *endo-*, *meso-*, and *derm-*. Underline the one that can also be used as a suffix.

3. Contrast some fresh, fertile eggs from a breeder, or from a health food store, with some ordinary grocery store eggs. Prepare a report, with display, indicating differences in appearance, price, and purported food value.

4. From an experienced cook, discover the unique culinary values of egg yolk and egg white. Write out some recipes illustrating these values.

◆ **Suggestions for Multimedia Resources:**[**]

1. "The Chick Embryo Microslides." *Ward's Natural Science.* To order a set of microslides for a reason-

able price, visit <http://wardsci.com/product.asp?pn=269017&sid=OVERTURE&EID=OV002657>.[***]

2. "Embryology: The 21-day Chick Lifecycle." 1998. University of Illinois at Urbana-Champaign Web site. Visit their interactive site at <http://chickscope.itg.uiuc.edu/explore/embryology/>; click on each day of the lifecycle for more information.

3. "Virtual Hatch Project." *4-H Virtual Farm.* Visit their Web site at <http://www.ext.vt.edu/resources/4h/virtualfarm/poultry/poultry_incubation.html>. This site is designed to help the student obtain a better understanding of incubation and embryonic development.

4. *The Multi-Dimensional Human Embryo Atlas.* Visit their Web site at <http://embryo.soad.umich.edu/carnStages/carnStages.html>. This site provides a three-dimensional image reference of the human embryo based on magnetic resonance imaging.

5. "Model Organism: Zebrafish." *Microscope Imaging Station.* Visit their Web site at <http://www.exploratorium.edu/imaging_station/zfish.html>. This site tracks the first 48 hours of zebrafish development, from a single cell to fry.

◆ **Suggestions for Supplementary Reading:**[**]

1. Butcher, Gary D., and Richard Miles. "Concepts of Eggshell Quality." *University of Florida, Institute of Food and Agricultural Sciences.* To read this article online, visit <http://edis.ifas.ufl.edu/VM013>.

2. Wineland, Michael J. "Specific Gravity Testing For Eggshell Quality." *Cedar Grove Farms.* To read this article online, visit <http://members.aol.com/CGFARMS/specificgravity.html>. Specific gravity determinations establish eggshell thickness and, therefore, eggshell quality.

◆ **Answers to Questions**

Growth of the Embryo (Text pages 97–98)
➢ **Questions: Growth of the Embryo**

1. *Parthenogenesis* is the development of an unfertilized egg. This occurs in various types of organisms: drones (males) of bees, female aphids, turkeys, sea urchins, frogs, and chicks.

2. The germ layers in order of their formation are as follows: the blastula stage forms the blastoderm; and the gastrula stage forms the ectoderm, mesoderm, and endoderm.

3. The differences between the development of a frog and that of a chick are as follows: in frogs, where there is a moderate amount of yolk, cleavage is retarded in the yolk so that it occurs more rapidly where there is no yolk (cells not containing

---

[*] For a free copy of *Killer Angel,* search for the title on the Internet or visit <**http://freebooks.entrewave.com/freebooks/docs/39ba_47e.htm**>; you may also download a PDF version.

[**] The listing of these suggestions does not necessarily imply endorsement of content.

[***] You may also want to browse their online catalog at <http://wardsci.com/category.asp_Q_c_E_620_A_Biology>.

yolk are smaller and more numerous); in chicks, where there is much yolk, cleavage cannot penetrate the yolk, and a raft of cells develops on the surface of the yolk.

In the frog, a restricted, slit-shaped area appears (*blastopore*). As the cells increase in number, those near the dorsal margin of the blastopore migrate inside. The cavity inside becomes a pocket partly lined with cells and partly by yolk. In chick embryos, the layer of cells may rise somewhat above the yolk, and cells grow under by a flowing motion rather than sinking in. The blastula stage is not a hollow ball as in the frog, but the blastoderm is still spread out on the yolk.

➤ Taking it Further: Growth of the Embryo

1. *Answers may vary. Using outside resources, the student is asked to draw different developmental stages of a chicken from fertilization to hatching.*

2. *Answers will vary. The student is asked to write an essay on the law of recapitulation, including supporting and opposing evidence for it.*

Experimental Embryology (Text page 99)

➤ Questions: Experimental Embryology

1. The following table shows the scientists and their contributions to the study of embryology:

Scientist	Contribution
Tichmiroff	induced parthenogenesis[a] by subjecting the eggs of silkworm moths to friction
William R. Breneman	questioned the "law" of recapitulation because it is inapplicable in many instances
Hans Dreisch[b]	separated the cells of sea urchin embryos after one and two cell divisions; each cell produced a complete embryo
Hans Spemann	discovered that if a portion of the dorsal (back) lip of the blastopore of a salamander embryo is transplanted to a region near the ventral lip of the blastopore, it induces the formation of another embryo on the ventral surface (abdomen) of the host embryo
Ernst Haeckel	Haeckel's "theory of recapitulation" (now debunked) stated that the embryonic development of a species would yield clues of its ancestry and development from organisms of a different sort

a. Parthenogenesis results in certain species naturally, but it may also be induced by artificial, external stimuli—either physical or chemical.

b. Wilhelm Roux, co-founder of **experimental embryology**, along with Hans Dreisch

2. Dreisch's assumption that all parts of the embryo have the same potentialities was faulty because not all eggs behave like the ones Dreisch studied.

3. Transplants of certain parts of the developing embryo show that some of them also act as organizers. For example, the area of a young salamander embryo that is normally destined to become a limb may be transplanted to a place on the body. Up until a certain time in the life of the embryo, this area will then develop into a normal part of the body at the new location instead of into a limb. But after this time, the result of such a transplant will produce a limb at the unusual place.

➤ Taking It Further: Experimental Embryology

1. *Answers will vary. The student is asked to find articles on cloning research and write a summary of the findings.*

2. *Answers will vary. The student is asked to write an essay, discussing the moral questions behind human cloning.*

➤ Questions: Chapter Review

1. The mesoderm forms muscles, bones, lining of body cavities, outer membrane of inner organs, inner layer of skin—the larger part of the body.

2. The embryo must have nourishment. The first cells to be formed lie next to the yolk and are able to digest the yolk and use it as food. But as the embryo grows, most cells are at a distance from this source of food. Nourishment is given to these cells by the blood, which the heart causes to circulate. Of course, such development is based upon good planning.

3. Long, slender fibers grow out from the cells which compose the brain and spinal cord. These fibers lie side by side and they, with their coverings, are called nerves.

4. An animal which develops by parthenogenesis has no father.

5. It is hard to be sure which seeds have developed by parthenogenesis and which by union with pollen. If all the seeds developed by parthenogenesis, there would be no ratio, but all would have the genotype of the mother.

6. The blastocoel of the frog is a sphere, while that of the chick is a thin area between the embryo and the yolk.

7. If the sea urchin were preformed in the egg and the embryo were divided at the two-cell stage, each cell would either fail to develop or would form half of a sea urchin.

8. An embryo is a living thing and needs coordination of parts. Another reason for early develop-

ment of the brain is that this organ is so complex that its formation needs time and must begin early.

◆ Think-Session Guide for Unit 3

The Continuity of Life

Subject: Cellular components

Purpose: To question how things got here

a. **To the student:** At a university hospital, a team of research physicians developed a technique of microsurgery in which they can operate on a single cell. An experiment was performed in which the nucleus was removed from cells *in vitro* to see what would happen. These data were recorded:

Number of cells *with nuclei* studied: 802

Number of cells *without nuclei* studied: 1093

Number of Cells Surviving After	With Nuclei	Without Nuclei
1 day	513	947
2 days	489	763
3 days	427	340
4 days	778	19
7 days	749	0
14 days	660	0
30 days	572	0

Assume that you had to make an interpretation at this point. What would you say about the importance of the nucleus? Why do the number of cells with nuclei increase and then decrease? Do you feel confident in your interpretation? Why?

Teacher: These questions are designed to force an examination and an interpretation. Of course, not a great deal should be said after one experiment. However, this could be a good place to discuss "sufficiency of data."

b. **To the student:** One of the newer members of the research team found a report in the literature that red blood cells exist for over 30 days without a nucleus. How might this influence your interpretation?

Teacher: Red blood cells do have a limited survival time which may be longer than the time of this experiment. Also, this was done *in vitro*. RBCs *in vitro* may have an environmental advantage.

c. **To the student:** What information would you like to have about this experiment to give you more confidence in drawing a conclusion?

Teacher: The student should ask about techniques. Could the operation itself cause the death? Are cells (living things) better able to survive in nature than under artificial conditions? Does the nucleus have direct or indirect control over cell activity? Where did the cell get its information?

UNIT 4
The World of Living Things

CHAPTER 9

Classification of Organisms
(Text pages 103-108)

Development of Classification
The System of Classification
Uncertainties in Classification

◆ Suggestions for Motivation or Enrichment:

1. Compile a list of at least ten different ways to classify people. Evaluate the scope of each classification.

2. Prepare a brief report on Theophrastus, the Father of Botany.

3. Research a general list of catlike animals and add any specific types that you know. Discuss the limits of possible interbreeding.

4. Divide a group of students into two panels to present the pros and cons of interracial marriage.

◆ Suggestions for Multimedia Resources:*

1. Conrad, Jim. "Names and Classification." *Backyard Nature.* To access this Web page, visit <http://www.backyardnature.net/names.htm>. This Web page discusses scientific names and explains why the classification of plants and animals is interesting. This site is evolutionary in orientation.

2. *The Five Kingdoms of Life.* Fogware Publishing. Visit their Web site at <http://www.fogwarepublishing.com/>. CD-ROM, Windows OS only. It explains the reason for the five-kingdom system and the criteria for determining the division; examines the life cycle and ecology of each kingdom.

3. Myers, Phil. *Animal Diversity Web (ADW).* To access this Web site database, visit <http://animaldiversity.ummz.umich.edu/site/index.html>. *ADW* is an online database of animal natural history, distribution, classification, and conservation biology at the University of Michigan. You may also want to visit the University of Michigan's Museum of Zoology site at <http://www.ummz.lsa.umich.edu/>.

◆ Suggestions for Supplementary Reading:*

1. Oard, Michael. "The Confusion of Elephant and Mammoth Classification." Appendix 1 from *Frozen in Time: The Woolly Mammoth, the Ice Age, and the Bible*, October 2004. To read this article online,

visit <http://www.answersingenesis.org/home/area/fit/appendix1.asp>. The author discusses problems with classifying these creatures.

2. Lamont, Ann. "John Ray: Founder of Biology and Devout Christian." *Creation*, 21: 151–53. To read this article online, visit <http://www.answersingenesis.org/creation/v21/i1/ray.asp>.

◆ Answers to Questions

Development of Classification (Text page 104)

➤ **Questions: Development of Classification**

1. It is important to have a universal system of classification because, without any grouping, the study of organisms would be confusing and difficult. It also allows anyone who reads biological reports on new knowledge a way to understand what is being described.

2. Some of the early methods of classification are as follows:

Scientist	Classification System
Ancients	no systematic way
Aristotle	**plants:** soft-stemmed (*herbs*); woody-stemmed (*shrubs*); and single-stemmed, trunks (*trees*) **animals:** land, water, and air dwellers
Other systems	strictly environmental or ecological
Theophrastus	studied the structure of the stems and leaves of **plants** and grouped them into families on the basis of the likenesses
17th & 18th centuries	long lists of known **plants** (*herbals*) and **animals** (*bestiaries*)
John Ray	presented the first clear concept of species, defined as offspring of similar parents; also addressed issues in physiology
Carolus Linnaeus	laid the foundation for an orderly system in the study of nature, focusing almost exclusively on classification

3. John Ray defined the term *species* as "offspring of similar parents."

➤ **Taking It Further: Development of Classification**

1. *Answers may vary. The student is asked to look up one of the scientists mentioned in this section and list the major works of his career.*

The System of Classification (Text page 105)

➤ **Questions: The System of Classification**

1. The different levels of classification from the broadest to the most specific are as follows: Kingdom, Phylum, Class, Order, Family, Genus, Species.[*]

2. The classification of dinosaurs:

Kingdom	Animalia (*animals*)
Phylum	Chordata (*with hollow nerve chord ending in a brain*)
Class	Archosauria (*diapsids with socket-set teeth, etc.*)
Subclass	Ornithodira (*dinosaurs and pterosaurs*)
Superorder	Dinosauria (*dinosaurs*)
Order	Saurischia and Ornithischia (*based on hip structure*)

Dinosaurs are divided into two orders based on their hip (*pelvic*) structure: the "lizard-hipped" or Saurischian dinosaurs and the "bird-hipped" or Ornithischian dinosaurs.

Teacher: Note that various phyla, classes, and orders of dinosaurs are not included at the end of this teacher's manual. The student may consult other classification systems to answer this problem.

➤ **Taking It Further: The System of Classification**

1. *The student is asked to give the seven classification levels (including any sublevels) for a horse. The student will have to use an encyclopedia, the Internet, or some other source to answer this problem.*

Kingdom	Animalia (*animals*)
Phylum	Chordata (*with hollow nerve chord ending in a brain*)
Class	Mammalia (*warm-blooded, mammary glands, live young suckled, presence of three middle ear bones, hair, diaphram, four-chambered heart, and large cereberal cortex*)
Order	Perissodactyla (*non-ruminant, herbivorous, odd-toed ungulate [i.e., hoofed mammal]—horses, rhinos, and tapirs*)
Family	Equidae (*horse family*)
Genus	Equus (*asses, horses, and zebras*)
Species	Equus caballus (*longer mane and tail, larger hoofs, more arched neck, smaller head, shorter ears*)

---

[*] Some add **Domain** as a level before Kingdom. The three Domains are Bacteria, Archaea, and Eukarya.

Uncertainties in Classification (Text page 108)

➤ **Questions: Uncertainties in Classification**

1. *Examples may vary. The student is asked to name the five kingdoms and give examples of each.*

Kingdom	Examples
Animalia	mammals, reptiles, birds, etc.
Plantae	herbs, shrubs, trees, etc.
Protista	algae (all kinds except for blue-green algae), dinoflagellates, diatoms, lichens,[a] etc.
Monera	bacteria,[b] blue-green algae, and mycoplasmas
Fungi[c]	slime molds, mushrooms, fungus, yeasts, rusts, smuts, etc.

a. a plant structure composed of an alga and a fungus growing together in a symbiotic relationship
b. single-celled prokaryotes, classified into two main branches, the Archaebacteria and the Eubacteria, each of which have several phyla
c. "plants" without chlorophyll

2. The difference between observable facts and theories is that *facts* are subject to experimental or observable verification, but *theories* are based on philosophical assumption, conclusion, or prediction regarding those facts.

3. There are more ways than one to classify the natural world because some one-celled organisms are difficult to classify as plant or animal. For instance, *Euglena* is an example of a microscopic organism that behaves like an animal in some ways, yet has chlorophyll that manufactures nutrients like a plant. Any system of classification involves a certain amount of arbitrariness.[**]

➤ **Taking It Further: Uncertainties in Classification**

1. *Answers may vary. The student is asked to identify the role of the arbitrary system of classification in supporting different theories of origin.*

➤ **Questions: Chapter Review**

1. In taxonomy, a class is a group of orders which have certain likenesses. A family is a group of genera which have certain likenesses.

2. Descriptions of plants or animals, if carefully made, are long and cumbersome. If a description is shortened, it is not exact. Another person, in describing this organism, would not choose the same parts to describe, with the result that it would be considered another organism. A name may be somewhat

---

[**] A complicating factor in classifying one-celled organisms is evolutionary conjecture. Evolutionists believe that these organisms are the common ancestors of both plants and animals because some life forms seem to combine both plant and animal characteristics. Yet, due to the fact that there are so few fossils, such hypothetical relationships are difficult to establish.

descriptive, but the real need is that it be a designation.

3. *First*, an alphabetical arrangement of names would not indicate which ones are somewhat alike. *Second*, if they were grouped according to likeness (e.g., warm-blooded, four-legged, furry animals) and then alphabetized, it still would not indicate the nature or inclusiveness of their subgroups. For example, a mole would be alphabetically listed close to a mustang, but they are still very different creatures (note their respective classifications).

4. In order to be used in various countries, names must be from one language. Latin, at one time, was used by more scholars than any other language; therefore it was selected as the language to be used for scientific terminology. If names are not in Latin, they are given latinized endings.

5. In a pure line, there is so much homogeneity that the genes in one plant are identical to the genes of any other plant. The same could be said of a clone; but it may be heterozygous, while a pure line is homozygous and so will not change in future generations except by mutation. A species does not have this degree of likeness among its members. Someone who is well acquainted with the organisms in question decides the limits of a species.

6. Genes normally occur in pairs, one from the father, the other from the mother. A dominant gene expresses itself by developing its trait and so does a recessive gene, but only when the dominant gene is absent. When dominant and recessive genes are together, the latter are latent and unobserved; sometimes, for man, for generations. Then, when the dominant gene is absent in a generation, the recessive gene expresses itself by developing its trait.

7. A race of plants or animals does not differ so much from other races as species differ from each other. Furthermore, it, along with varieties and breeds, can mate with individuals of other groups and so lose its identity in time. A geographical race is kept from mating with others by dwelling in a separate place.

◆ Think-Session Guide for Unit 4

The World of Living Things

Subject: Numerical progressives

Purpose: To try to find patterns in a "random" list

a. **To the student:** One of your first assignments on a new job is to make sense out of the following list:

*"A*020819808*B*003910562*C*012019374*D*020860 687*B*023919805*Z*141549506*"*

Teacher: After a brief time, suggest that the letters may stand for names. This may lead to thinking of addresses. Suggest that the last **5 digits** may be a *zip code*. It should be easy to convince the student that the other **4 digits** are *house numbers* since **A** and **D** are the same.

---

b. **To the student:** What is the relationship between the two *Bs*?

Teacher: Obviously, more information is needed. Point out that this is not unlike the problem facing taxonomists who try to catalog something they did not create.

UNIT 5
Small Plants and Little Animals

CHAPTER 10
Kingdom Fungi: Non-green Plants
(Text pages 111-118)

Classification of Fungi

The Molds

Other Fungi

Benefits from Fungi

◆ Suggestions for Motivation or Enrichment:

1. Discuss with others the experiences with such common diseases as athlete's foot, ringworm, and valley fever[*] (*Coccidioidomycosis*).

2. Separate the word *antibiotic* into its two main parts, and suggest the logic by which it was formed. Make a list of common antibiotics and the ailments which they tend to prevent, inhibit, or destroy.

3. Gently break apart some very moldy bread, examine it with a hand lens, and sketch (with labels) a section of mold with spores and connectors.

4. Research and prepare a report on cheese making. If possible, include samples in your report.

◆ Suggestions for Multimedia Resources:[**]

1. Microbe World is an official Web site for the *American Society for Microbiology*. To access this Web site, visit <http://www.microbeworld.org/>. This interactive site provides basic information for those interested in learning more about microorganisms; it is oriented towards evolution.

◆ Suggestions for Supplementary Reading:[**]

1. McQueen, Rodney, and David Catchpoole. "The 'Fungus' That 'Walks.'" *Creation*, 22: 349–51. To read this article online, visit the *Answers in Genesis* Web site at <http://www.answersingenesis.org/creation/v22/i3/fungus.asp>.

◆ Answers to Questions

Classification of Fungi (Text page 112)

➤ **Questions: Classification of Fungi**

1. It was difficult to classify fungi because they are "plants" without chlorophyll and are unable to manufacture food. For example, during a part of its life cycle, a **slime mold** resembles a one-celled animal and might be considered an *animal*; but at another stage, it produces spores in structures, characteristic of molds, and is definitely *plantlike*.

Also, a **lichen** is a fungus and an alga growing in a symbiotic relationship, and it is usually classified with algae, under Kingdom Protista. (*See "Lichens: Two Plants in One" in chapter 12.*)

2. A *spore* is a single cell that can grow into a mature fungal plant. It is different from a *seed* in that it contains no embryo plant and very little stored food. Since it lacks these features, a spore cannot grow unless conditions are very favorable. In addition to this asexual method, some fungi reproduce sexually by the union of two gametes to form a zygote.

3. The main difference among the phyla of this kingdom is the type of reproduction used—*conjugation fungi, sac fungi, club fungi,* and *imperfect fungi.*[***]

Teacher: In the first printing of the textbook, at the bottom of page 111, the text states "The members of the <u>fourth phylum</u> in the Fungi kingdom are separated by the type of reproduction used...." It should read "The members of the <u>four phyla</u> in the Fungi kingdom are separated by the type of reproduction used...." (*This was corrected in the 2006 printing.*)

➤ **Taking it Further: Classification of Fungi**

1. *Answers may vary. The student is asked to list several positive and negative effects of fungi.*

The Molds (Text page 113)

➤ **Questions: The Molds**

1. The conjugating (or algal) fungi belong to the Phylum Zygomycota. (*See the appendix at the back of this teacher's manual.*)

2. Besides making cheese, *Penicillium* was the first antibiotic discovered and was the one most widely used until resistant strains of bacteria became more abundant.

3. *Parasites* obtain food from living hosts, and *saprophytes* obtain food from nonliving organic matter.

➤ **Taking It Further: The Molds**

1. *Answers may vary. The student is asked to find an example of a living mold and, by its color and location, try to determine its genus.*

---

[*] This fungus is commonly found in the soil of the southwestern United States, Mexico, and parts of Central and South America.

[**] The listing of these suggestions does not necessarily imply endorsement of content.

[***] Some add a fifth phylum, **Oomycota** (Protistalike Fungi), which includes the water molds that form on fish, on matter in water, etc.

Other Fungi (Text page 116)

➤ Questions: Other Fungi

1. The *sac fungi* belong to the Phylum Ascomycota, *club fungi* belong to the Phylum Basidiomycota, and *imperfect fungi* belong to the Phylum Deuteromycota. (*See the appendix at the end of this teacher's manual.*)

2. A *mycosis* is a fungal infection that attacks man and is extremely difficult to cure and often remains for many years.

3. Rusts require two hosts. For example, wheat rust attacks its **first host** (wheat) during the time it is producing seeds. Late in the summer, the fungus enters a second stage in which it forms black spores on the dry stalks left in the fields. These spores then are blown by the wind and fall on its **second host** (barberry bushes), where they attack the leaves in the spring.

➤ Taking It Further: Other Fungi

1. *Answers may vary. The student is asked to give an example of each type of fungi and describe its method of reproduction.*

Benefits From Fungi (Text page 118)

➤ Questions: Benefits From Fungi

1. *Answers may vary.* Benefits of fungi are as follows:
   - *For healing*, such as penicillin—an antibiotic used to fight influenza bacteria
   - *For food*, such as edible mushrooms, truffles, and cheeses (produced by the action of molds on milk)
   - *For decomposition*, such as those which break down dead plants and animals—recycling chemical elements in their cytoplasm
   - *For root absorption*, such as mycorrhizae that help pine trees
   - *For insect control*, such as several species of fungi that parasitize insects

2. Decomposers, mostly bacteria and fungi, are *heterotrophs* (organisms that require complex organic compounds of nitrogen and carbon for metabolic synthesis) that obtain food by breaking down substances in dead protoplasm. They are so important because without them the earth would be littered with dead bodies.

3. Fungi have medicinal use as an antibiotic, such as penicillin that fights influenza bacteria. In the production of food, some fungi are edible, such as mushrooms and truffles; baker's yeast is used to make bread rise and to brew beer; and some fungi (i.e., molds) are used in the production of cheeses.

➤ Taking It Further: Benefits From Fungi

1. *Answers may vary. The student is asked to find an article on the use of any fungus as an insecticide and summarize the findings.*

➤ Questions: Chapter Review

1. Many molds live on organic substance which is not living; therefore, they are called *saprophytes*. Quite a number of molds live on living things, but most of the ones you see are saprophytes.

2. The essential structures of a fungus are fibers called *hyphae*, *spores*, and some structure on which or in which the spores are found. Hyphae may be given different names according to their position.

3. A yeast cell produces a mere protrusion called a bud, which grows and becomes a full-sized cell. *Penicillium* forms a branched stalk. On the ends of these branches grow tiny cells called spores that have the ability to grow into complete mold plants.

4. Since the Japanese barberry is harmless and the common barberry harbors wheat rust, it is important that they be named correctly and that persons learn to recognize them.

5. The alcohol formed by yeast is very volatile and so is dissipated in the oven.

6. Athlete's foot is a parasitic mold. A person should avoid walking barefoot in areas where there may be contamination.

7. Darkness is neither an advantage nor a disadvantage. An even temperature helps mushrooms and an even degree of moisture in the air also helps a great deal. However, a sudden drying of the air would kill them.

8. Mushrooms do not live on mineral soil but on plant or animal material. Garden soil has but little of plant or animal flesh or refuse.

9. If air can circulate between layers of boards, it keeps the boards so dry that molds cannot grow on them. Molds are the chief agents of the decay of wood.

10. Some might say that Sir Alexander Fleming's discovery was "chance"; but, under the direction of God's providence, penicillium spores were allowed to get into the culture dishes of bacteria. Perhaps the wind blew in some spores when someone opened the dishes. But the discovery of what the spores had done was due to the careful training of Alexander Fleming.*

---

* Helpful discoveries always require training and research that a so-called "chance" event may be recognized and developed.

CHAPTER 11
Viruses, Bacteria, and Other Forms
(Text pages 121-132)

Viruses
Bacteria
Other Forms of Microorganisms
Control of Microorganisms

◆ Suggestions for Motivation or Enrichment:

1. Consult a recipe book or online source for information about making **sauerkraut** (chopped or shredded cabbage that is salted and fermented in its own juice). Prepare a display that indicates the role of bacteria in the process.

2. Prepare a demonstration of linear measurement. Mark on a stick the following measurements: 1 meter, 1 yard, 1 inch, 1 centimeter. Identify some known object of 1 millimeter size, and discuss the micron as 1/1000 of that size.

3. Compile a list of minute structures or creatures that are not visible to the unaided eye.

4. Follow the advice of a plant nurseryman or botany teacher in obtaining specimens of local plants that have a viral disease. Make a display, indicating the name of each disease and its treatment.

◆ Suggestions for Multimedia Resources:*

1. "Microbe Zoo." *Digital Learning Center for Microbial Ecology.* Visit this Michigan State University site at <http://commtechlab.msu.edu/sites/dlc-me/zoo/>. This is an interactive site about bacteria.

2. "The Picture Gallery." *Protist Image Data.* To access this University of Montreal Web site, visit <http://megasun.bch.umontreal.ca/protists/gallery.html>. This site provides pictures and information on selected genera of algae and protozoa, and resources in protistology and related fields—microbiology, mycology, phycology, and protozoology.

3. "The Hidden Killers: Deadly Viruses." To access this interactive site, visit <http://library.thinkquest.org/23054/gather/index.shtml?tqskip1=1>. This Web site provides information about viruses, including the human immune system, specific virus profiles, and the research of epidemiologists.

◆ Suggestions for Supplementary Reading:*

1. Mayer, Gene. November 1, 2004. "Immunoglobulins—Structure and Function." *Microbiology and Immunology On-line.* To access this article, visit the University of South Carolina site at <http://path-

---

* The listing of these suggestions does not necessarily imply endorsement of content.

micro.med.sc.edu/mayer/IgStruct2000.htm>. It outlines the function and structure of antibodies.

2. "Antifungal Agents: Introduction." *The Merck Veterinary Manual.* To read this information online, visit <http://www.merckvetmanual.com/mvm/index.jsp?cfile=htm/bc/191300.htm>.

3. "Cholera." From *Wikipedia.* Modified September 5, 2006. To read this article online, visit <http://en.wikipedia.org/wiki/Cholera>.

4. Green, Steven N. 1991. "Herpes Viruses." *Eclectic Dentistry.* To read this online article, visit <http://www.sngreen.com/book/12.htm>.

◆ Answers to Questions

Viruses (Text page 125)
➤ Questions: Viruses

1. The three types of organisms that are preyed upon by viruses are (1) bacteria, in which bacteriophage live; (2) plants, in which viruses attack plant cells; and (3) humans and animals, in which viruses attack human and animal cells.

2. The four different types of viral infection are as follows: *lytic, latent, persistent,* and *transforming.*

Note: The virus's reproductive cycle involves the following stages: (1) the virus attaches to the cell through the use of the tail fibers, and the virus releases its nucleic acid into the cell; (2) once inside the cell, the nucleic acid of the virus takes over the chemical functions of the cell and uses the cell's organelles to reproduce the different components of the virus; (3) the individual components of the virus assemble, and then the viruses are released according to the type of infection caused by the virus.

- *Lytic infections* go through all the above steps. Once the viruses have assembled, the cell *lyses* (bursts or splits), releasing the viruses to the environment. A lytic infection ends in cell death.
- A *latent infection* goes through the first two steps, but once the nucleic acid enters the nucleus of the cell, it inserts itself into the host cell's DNA. The latent virus's genetic code is replicated with each mitotic division of the cell and remains inactive until an outside stimulus causes all the infected cells to finish the virus's reproductive cycle.
- *Persistent infections* follow the normal reproductive cycle; but, unlike the lytic cycle, the host cell is not destroyed in the process. As the viruses are assembled, they are released through the cell membrane.
- *Transforming infections* are similar to latent infections in that the virus inserts itself into the host's DNA. However, instead of remaining inactive, it causes the host cell to reproduce abnormal cells. Each time the cell divides, more abnormal cells are produced. These abnormal cells have been shown to cause *cancer.*

3. Interferons are important in the study of disease because they prevent a second infection and may have something to do with the recovery of an

infected individual. Interferons may also prevent the host from being infected by other kinds of viruses. Some scientists believe that the interferon from intestinal virus infection prevents polio.

➤ **Taking It Further: Viruses**

1. *Answers may vary. Using the characteristics of living material, the student is asked to discuss the nature, either living or non-living, of viruses.*

2. *Answers may vary. The student is asked to write an essay discussing why proving that viruses are an intermediate stage between life and non-life would be of value to evolutionists.*

Bacteria *(Text page 128–129)*

➤ **Questions: Bacteria**

1. The three main shapes of bacteria are as follows:
   - *Spirillum*, a cell shaped like a corkscrew or curved rod
   - *Bacillus*, a little, straight rod; and *streptobacilli*, chains of rods joined end to end
   - *Coccus* is a tiny sphere that may occur singly; in pairs, *diplococcus*; in chains, *streptococcus*; or in branched chains, *staphylococcus*

2. Environmental factors that are important to keep in mind when growing a bacterial culture are as follows: suitable temperature (highly variable for different forms), moisture, sunlight (for *photosynthetic bacteria*), nutrition, and oxygen.

3. The three different types of genetic recombination in bacteria are *conjugation*, *transformation*, and *transduction*. In conjugation, the two cells join and form a strand of cytoplasm between. Transformation occurs when bacteria die and release DNA into the environment, which is then absorbed by other bacteria. Transduction is facilitated by viruses, which transfer DNA from one bacterium to another.

4. Bacteria are beneficial to mankind in the following ways: *Lactobacillus* is used to form lactic acid, which is used to produce many milk products (as butter, buttermilk, and sour cream); *Lactobacillus bulgaricus* and *Streptococcus thermophilus* are added to milk and milk solids to make yogurt; the action of bacteria on milk also produces cheeses (distinctive flavors of cheeses result from the kinds of bacteria used). In addition, acetic acid bacteria are used to make vinegar; the action of *Lactobacillus* on cabbage yields sauerkraut; the action of bacteria on the soft parts of the stems of flax form linen fibers; bacteria is used to cure tobacco leaves; and bacteria is used to ferment chopped corn, oats, or clover to make silage.

5. Bacteria perform various functions in the ecological world—as *decomposers*, they return chemical substances to the soil; as *food*, they are eaten by protozoa; and as *pathogens*, they cause plant diseases, which adversely affect the farmers' crops.

➤ **Taking It Further: Bacteria**

1. *Answers may vary. The student is asked to find several pictures of different bacteria and summarize their natural environments, and give the proper name and type (based on shape) of the bacteria.*

2. *Answers may vary. The student is asked to look up the Latin roots of the following words: Archaebacteria, Eubacteria, prokaryotic; he is also asked to give their meanings.* **Archaebacteria** are part of a diverse group of bacteria (prokaryotes); the prefix *archae-* means "ancient"; so they are ancient forms thought to have evolved separately from other bacteria. The prefix *eu-* means "true." The so-called **Eubacteria** or "true bacteria" are all the organisms traditionally known as *bacteria*.

   Prokaryotic refers to organisms of the kingdom Monera (bacteria and blue-green algae), characterized by the absence of a distinct, membrane-bound nucleus. The prefix *pro-* means "earlier than, prior to"; *kary-* comes from the Greek word *karyon* which means "nut" (i.e., nucleus); the suffix *-otic* means "of, relating to." Thus *prokaryotic* refers to cells of organisms that are "earlier" than cells with a membrane-bound nucleus.

Other Forms of Microorganisms *(Text page 130)*

➤ **Questions: Other Forms of Microorganisms**

1. A *vector* is a living thing that carries disease-causing microorganisms from one host to another.

2. Rickettsiae and viruses are similar in that they both require a living host to survive; they are also hard to study, since they both require a living culture medium, such as a chicken embryo.

3. The **bacterium** (*R. prowazekii*) and the **vector** (*ticks, fleas,* and *rats*) are the two parasites that are responsible for the spread of typhus.

➤ **Taking It Further: Other Forms of Microorganisms**

1. *Answers may vary. The student is asked to explain why increasing sanitation during the war proved to be an effective method of controlling typhus and to give examples supporting his reasoning.*

Control of Microorganisms *(Text page 132)*

➤ **Questions: Control of Microorganisms**

1. An **acquired immunity** is gained during an individual's lifetime and cannot be passed on to its offspring; there are two types—active and passive. A **natural immunity** is present in the individual at birth and is not artificially acquired.

2. There are five types of **antibodies**—*antitoxins*, which neutralize the toxins of the microorganism; *agglutinins*, which cause a clumping of bacteria; *precipitins*, which make the toxins insoluble and precipitate them out; *lysins*, which dissolve bacterial cells; and *opsonins*, which make the

microorganisms more susceptible to attack by the phagocytes.

3. An *active immunity* is acquired by having a specific disease. A person who has had chickenpox, polio, or smallpox will probably never contract the disease again; also, a related disease (e.g., cowpox) may protect against a more violent disease (smallpox infection). A *passive immunity* is acquired from another individual or organism. For example, antibodies pass through the placenta into the child's bloodstream; for six months after birth, the child is usually immune to the diseases to which the mother was immune. Also, antibodies can be passed to the baby through breast-feeding and remain effective up to eighteen months after the mother stops.[*]

4. *Chemotherapy* is treatment of infectious diseases with chemicals that harm the microbes without harming the patient.

➤ **Taking It Further: Control of Microorganisms**

1. *Answers may vary. The student is asked to determine which of the following would have the greatest chance of seriously harming a person: a single molecule of botulinum exotoxin (soluble poisonous substance produced by* Clostridium botulinum*), a single virus of smallpox, or a single molecule of a highly radioactive substance, and why.*

➤ **Questions: Chapter Review**

1. Bacteria occur in three shapes or forms as follows: (1) *spirillum* is a cell shaped like a corkscrew or curved rod; (2) *bacillus* is a little, straight rod, and *streptobacilli* are chains of rods joined end to end; and (3) *coccus* is a tiny sphere that may occur singly, in pairs (*diplococcus*), in chains (*streptococcus*), or in branched chains (*staphylococcus*).

2. The reasons for thinking viruses to be living things are as follows:

   • Living (organic) things have definite size and shape; viruses, too, occur in definite sizes and shapes.
   • Living things are chemically active; viruses also are chemically active while inside a cell, though the activity is of a different nature from that of the cells.
   • Living things have a life span; and viruses also are active or "live" a brief time in cells, stimulating the production of other viruses.
   • Living things originate, grow, age, and die; viruses go through cycles but apparently do not age.
   • Living things are organized in units called cells; viruses also are in units, but not as complex as cells.
   • Living cells are composed of highly complex systems and chemicals; the core of a virus is nucleic acid, but the shell of a virus is composed of protein only, without the complexity of other compounds or integrated systems.

   • Living things require a source of energy (directly or indirectly from the sun); viruses appropriate the energy of the host cell (indirectly from the sun).
   • Living things are able to reproduce; viruses do not reproduce alone, but they cause the cell to produce more viruses so that a kind of reproduction does take place.
   • Living things respond to stimuli (light, heat, or chemicals); viruses also respond to stimuli (the presence of host cells).
   • Living things have a critical relationship to environment (temperature, light, moisture, and chemical environment) ; viruses are also affected by environment (colds are more common in the winter months, and polio occurs more often in the summer months).

Teacher: It is good mental training for the student to learn the facts given in the textbook and form his own conclusion as to whether viruses are living things. (Dr. Duane Gish,[**] a Creationist biochemist, stated, "A virus is as dead as a salt crystal.")

---

3. It has not been demonstrated that *anything* is an ancestor of other living things which are much more complex than they themselves are. Another difficulty in considering a virus to be the ancestor of plants is that the virus must be within the plant—a parasite, depending upon it. It could not be an ancestor of its host.

4. In soil we usually find more dead organic material than living things. Where this is true the saprophytic bacterium would be the more likely to survive.

5. **Purpose of formation:**

   • Spores of molds are formed for reproduction. Live spores act like seeds, forming new mold growths (colonies) when they find the right conditions.
   • Spores of bacteria, however, are not formed for reproduction but for carrying the bacterium through an unfavorable period. Such spores are not formed in large numbers in a place where they are easily carried away, as reproductive spores are, but in a cell, covered by a protective wall.

   Method of formation:

   • Blue-green molds reproduce by sending up spore-bearing stalks called *conidiophores* (fruiting structures) from which free spores (*conidia*) are released.
   • Spore formation is limited to certain bacteria and provides for survival in unfavorable conditions. The spores, called *endospores,* are composed of the cytoplasm drawn into a tiny ball or oval structure sur-

---

[*] For infants, the advantages of developing passive immunity through breast-feeding far outweigh the dangers that accompany active immunity, especially toxic vaccines.

[**] He has held key positions at Berkeley, Cornell University Medical College, and The Upjohn Company, where he collaborated with former Nobel Prize winners in various projects. His interest in the creation/evolution issue grew until, in 1971, he left The Upjohn Company to join the faculty at the newly established (1970) Christian Heritage College and its research division. In 1972, the latter changed its name to the Institute for Creation Research, and Dr. Gish has served as Associate Director and Vice President since that time.

rounded by a hard case. Endospores can survive drying as well as the temperature of boiling water. Spores found frozen in ice in Antarctica became active when placed in a favorable environment.

6. Active immunity is superior in that it lasts longer than passive immunity and is not dependent upon remedies which are hard to obtain. But it is not wise to be inoculated with the disease on purpose in order to be protected in the future by active immunity.

7. Vectors (living things that carry disease-causing microorganisms from one host to another) of rickettsial diseases are insects or animals, such as ticks and fleas. In the United States, rats serve as a reservoir of the disease. When a sick rat dies, the fleas immediately leave the body. Persons bitten by migrating fleas can get typhus if the rat was a carrier. Rocky Mountain spotted fever is carried from wild animals to man by the tick. In Europe, human body lice carry typhus from person to person.

8. Spores of bacteria are not killed by boiling water unless boiled three hours or placed in boiling water or steam under pressure. The pressure causes the temperature of boiling water to rise above the normal boiling point.

9. A disinfectant is usually a chemical placed on a surface or in the air to kill the germs of disease. An antibody is found in a living thing, where it works against foreign substances. An antibiotic is formed by some living thing, such as a mold, and hinders the growth of some bacterium or similar thing. A vaccine produces a disease in a weakened form, in order that the human body can overcome it by active immunity. These agents have in common the ability to work against disease.

CHAPTER 12
Algae
(Text pages 135-141)

Classification of Algae
Chlorophyta: Green Algae
Chrysophyta: Algae in Shells
Phaeophyta: Brown Algae
Rhodophyta: Red Algae
Dinoflagellates
Importance of Algae

◆ **Suggestions for Motivation or Enrichment:**

1. Research library reference materials for modern uses of kelp and of diatomaceous earth. Write a paragraph on each.

2. Check indices of news magazines for articles on the "Red Tide" around Florida, or of startling sur-

face appearances of luminescence in various areas of world oceans.

3. Trace a map section which includes the Sargasso Sea, label it, and add brief notes of interest about the sea.

4. From personal experience, or from inquiries, prepare an oral report on the seaweed (kelp) of some beach area. Include any information about kelp beds, kelp-cutters, types of unique plantlike structures on the kelp, and the problem for swimmers or boaters.

◆ **Suggestions for Multimedia Resources**[*]

1. Conrad, Jim. "Lichens." *Backyard Nature*. To access this Web site, visit <http://www.backyardnature.net/lichens.htm>. This site gives a brief explanation of lichens and offers links to differing forms of lichens and lichen-related Web sites; elsewhere, this site has references to evolution, although not this particular page.

2. "Lichenland: Fun with Lichens." To access this *Northwest Alliance for Computational Science & Engineering* Web site, visit <http://mgd.nacse.org/hyperSQL/lichenland/>. This site illustrates the cooperative relationship between fungi and algae.

◆ **Suggestions for Supplementary Reading:**[*]

1. Mahmoud, A. L. E., A. A. Issa, and M. H. Abd-alla. 1992. "Survival and Efficiency of N_2-Fixing Cyanobacteria in Soil under Water Stress." *Journal of Islamic Academy of Sciences*, Volume 5, No.4. To read this article online, visit the *JIAS* Web site at <http://www.medicaljournal-ias.org/5_4/Mahmoud.htm>. Survival of five genera of N_2-fixing cyanobacteria were studied under salt and drought stress in clay and sand soil.

2. "Blue-green Algae (*Cyanobacteria*) and Their Toxins." *Water Talk*, June 25, 2003. To read this article online, visit this *Health Canada Online* site at <http://www.hc-sc.gc.ca/ewh-semt/water-eau/drink-potab/cyanobacteria-cyanobacteries_e.html>. This document covers a wide range of topics related to cyanobacteria, their toxins, and human health: background, effects on humans and animals, drinking water, dialysis patients, recreational water, fish consumption, blue-green algal products.

3. Criqui, Nan. 1972. "Voyager: Giant Kelp." To read this article online, visit this *University of California, San Diego (UCSD)* at <http://aquarium.ucsd.edu/learning/learning_res/voyager/kelpvoyager/index.html#>. This interactive site gives basic information on the physical features, location, and current uses of giant kelp.

---
[*] The listing of these suggestions does not necessarily imply endorsement of content.

◆ Answers to Questions

Chlorophyta: Green Algae (Text page 137)

➤ **Questions: Chlorophyta**

1. The chloroplasts give this phylum its characteristic grass-green color.

2. Reproduction varies between asexual or sexual among the members of this phylum.

   - *Protococcus* reproduces asexually by fission (spontaneous division of the body into two or more parts, each of which grows into a complete organism).
   - *Chlamydomonas* reproduces sexually when joining gametes fuse to form a zygote (fertilized egg).
   - *Spirogyra* reproduce by conjugation, which takes place between two adjacent cells; a tube forms that connects the two cells, and the contents of one cell pass into the second cell—both asexual and sexual reproduction occur, but sexual reproduction occurs in response to unfavorable conditions.
   - *Ulothrix* reproduces asexually and sexually.
   - *Oedogonium* and other similar species reproduce when one filament produces two kinds of gametes; some cells produce sperm cells that are similar to zoospores of *Ulothrix* in appearance.

3. The *Chlamydomonas'* cell wall is composed of a protein and a carbohydrate instead of cellulose.

Chrysophyta, Phaeophyta, Rhodophyta, Dinoflagellates (Text page 138)

➤ **Questions: Chrysophyta, Phaeophyta, Rhodophyta, Dinoflagellates**

1. The typical color associated with each of the following phyla is as follows:

   - Chrysophyta—golden or yellowish green
   - Phaeophyta—brown
   - Rhodophyta—red
   - Dinoflagellates—red or pink

2. An *alternation of generations* is a type of life cycle in which the asexual reproductive stage alternates with the sexual reproductive stage.

3. *Answers may vary. The student is asked to name one positive and one negative ecological effect of the members of each group.*

   - **Diatom**—(+) supplies food to tiny animals and forms diatomaceous earth, which acts like a filter; (–) gives "fishy" taste to fish
   - **Brown algae**—(+) largest of all plants and the fastest grower; tremendous economic value (food and many ordinary products we use)
   - **Red algae**—(–) little known economic value; form enormous masses that have interfered with the movement of ships
   - *Fucus*—(no information given)
   - **Dinoflagellates**—(+) food for other organisms, some are *saprophytic* (consuming dead organic matter) and others are *phototropic* (help manufacture food in the sea); (–) forms "red tide,"
   which destroys millions of fish, and the dead fish washed up on the shore produce a stench as well as a health hazard

Importance of Algae (Text page 141)

➤ **Questions: Importance of Algae**

1. A *lichen* is a plant structure composed of an alga and a fungus growing together in a symbiotic relationship. The alga provides food for the fungus, and the fungus provides security and moisture for the alga.

2. In the Arctic, reindeer moss (a lichen) is an abundant source of food for reindeer. In the Arctic and Antarctic regions, diatoms feed the great whales. Small crustaceans, crab- and lobster-like animals, and other creatures eat the algae. Not only does algae form the base of the **food chain**, but also they may directly provide food for man. *Chlorella* (green algae) has been used experimentally as flour for baking and is a possible source of oxygen and food for astronauts. Red algae (common form called *dulse*) are used for food by Japanese, Hawaiian, English, and Scottish peoples; they are also used by the Japanese for thickenings in food.

3. An abundance of algae in a body of water (e.g., the "red tide" formed by *dinoflagellates*) is hazardous because it destroys millions of fish and, subsequently, the dead fish wash up on the shore, producing a stench as well as a health hazard. Also, a great quantity of algae may die from overcrowding, and the decaying plants produce poisons that destroy fish or livestock that drink the water. In large numbers, the plants themselves seem to be poisonous.

4. Nitrogen-fixing algae (*Nostoc* and *Anabaena*) and bacteria are very important because they are the only organisms that can form nitrogen compounds and fix them in the soil.

➤ **Taking It Further: The Importance of Algae**

1. *Answers may vary. The student is asked to explain how lichens might be used in widening our influence on our solar system.*

➤ **Questions: Chapter Review**

1. *Protococcus* grows most abundantly on the shaded side of the tree trunk, which usually is the north side. But consider where the tree is standing. Do not look at one which is shaded by other trees on one side.

2. Five methods of reproduction of algae are fission, accidental breaking of filaments, conjugation, formation of spores which are alike, and formation of sperm and eggs. The last is most like reproduction of land animals.

3. *The authors of the textbook consider classification a matter of convenience. Even on this basis, it would be*

well for all biologists to agree in classifying organisms, but unfortunately such agreement is not complete. It seems strange to classify plants as large as kelps among the Protista, but their reproduction is more like that of the algae than like that of any other plant.

4. In a lichen, the alga is held like a slave by the fungus, which uses part of the food which the alga produces by means of its chlorophyll. Yet, the fungus holds the alga in a position where it receives sunlight, and this position, along with the network of the fungus, holds rain water. The alga is either a sharecropper or a slave that is well treated.

5. If an Eskimo hunter killed a seal for his family, the links in the food chain through which this food came are as follows: algae in the water produced sugar, starch, and protein; the alga was eaten by a little animal (*such as* Cyclops—*see page 162 in text*); the *Cyclops* was eaten by a fish; and the fish, in turn, by a seal; finally, the Eskimo eats the seal.

6. The **direct method** is for people to eat algae (such as *Nostoc* and kelp). The **indirect method** is to care for a fish pond in such a way that the algae, which are food for small fish, will thrive. Or perhaps algae are eaten by tiny animals which the fish eat. Lastly, the fish are eaten by man.

7. Soil bacteria and molds break down dead plants and animals, changing them to humus. When mixed into a soil, humus makes it porous, holding some air which is needed by plant roots. Even if algae grow on the surface of humus, they do not exclude all the air.

8. *Answers may vary. The student is asked to name only one alga.*

   Nostoc and *Protococcus* become inactive when dry and resume activity when supplied with water. Thus, they are superior to large plants. However, no plant is superior to all others in every respect.

CHAPTER 13
One-celled Organisms: The Protozoa
Text pages 143–148

Description of Protozoa
Importance of Protozoa
Malaria and Quinine

◆ Suggestions for Motivation or Enrichment:

1. Presume that you are a hospital administrator in a metropolitan area faced with trying to pinpoint the source of an epidemic of amebic dysentery. List some key questions which you should ask each victim who arrives at your hospital.

2. Tie together the ends of an 8-inch string or soft wire and form it into a circle on paper laid over a hard surface. Then form a beginning pseudopodium from the circle and continue the formation as you move the ameba-figure across the paper. Sketch around each major move, and fill in arrows to show the flow of fluid substance into the pseudopodia.

3. Make a chart of malarial stages in human blood, based upon some illustration in a physiology reference book. If possible, discuss the victim's experiences with some medical person or with someone who has lived in the tropics.

4. Check a World Health Organization (WHO) bulletin for a graph or other announcement of WHO involvement in the care of protozoal infections. Prepare a report on the pertinent facts.

◆ Suggestions for Filmed Material:[*]

1. *God's Earth Team*/Video (Moody, 30 min.). A fast-paced journey that starts with protozoa and climbs to the top of the food chain, showing how all of God's creatures are interconnected

◆ Answers to Questions

Description of Protozoa (Text page 146)

➤ **Questions: Description of Protozoa**

1. The various forms of locomotion used by protozoans are pseudopods (Sarcodina), cilia (Ciliates), movement in blood of hosts (Sporozoa), and flagella (Mastigophora).

2. *Answers may vary. The student is asked to name a member of each phylum discussed above.*

   • Sarcodina—*Amoeba proteus, Entamoeba histolytica*
   • Ciliates—*Paramecium caudatum, Paramecium bursaria*
   • Sporozoa—*Plasmodium, Trypanosoma*
   • Mastigophora—*Euglena*

3. The movement of food through a paramecium occurs as follows:

   • The beating of the cilia and the forward movement of the creature sweep bacteria through the oral groove into the gullet.
   • Bacteria terminate in a gullet at the bottom, where a food vacuole forms.
   • A vacuole forms and eventually moves into the cytoplasm, carrying the bacteria.
   • The cytoplasm streams in a circular motion, carrying food vacuoles around the paramecium toward the anal pore.
   • The food is digested when the vacuole is joined by a lysosome, and indigestible material remains in the vacuoles.
   • These vacuoles (now waste vacuoles) can be seen in the cytoplasm.
   • The solid wastes are extruded through the anal pore located near the posterior end.[**]

---

[*] The listing of these suggestions does not necessarily imply endorsement of content.

4. The distinguishing characteristics of each of the phyla mentioned above are as follows:

- **Sarcodina**—masses of jelly with a single nucleus (in most species); flow of cytoplasm forms pseudopods, which are used for locomotion
- **Ciliates**—shaped like sole of shoe; locomotion by cilia, which cover the body; has two nuclei; attracted to slightly acidic areas; extrudes tiny barbs to avoid solid objects or for defense; two methods of reproduction (cell division and conjugation); *P. bursaria* is unique: does not eat, has zoochlorellae (tiny algae) that live in its tissue and produce food for both
- **Sporozoa**—no means of locomotion; parasitize most animal groups and other protozoa; nearly all cause disease (e.g., malaria—see pages 145–146 of the text for reproduction in hosts)
- **Mastigophora**—resembles both plants and animals; many, but not all, have chloroplasts that manufacture glucose; if light is scarce, they absorb food from the water; locomotion by flagella—*Euglena* also moves by changing body shape; *Euglena* also characterized by a red, light-sensitive eyespot

➤ **Taking It Further: Description of Protozoa**

1. *Answers may vary. The student is asked to describe how each of the four examples of protozoans above carry out the nine life processes (see p. 143). The nine life processes are* respiration, nutrition, circulation, response, support, reproduction, movement, growth, *and* excretion.

Importance of Protozoa (Text page 147)

➤ **Questions: Importance of Protozoa**

1. *Answers may vary. The student is asked to list the positive and negative* **ecological effects** *of these protozoans.*

- Hundreds of species of protozoa serve as food for other animals.
- *Paramecia* consume bacteria.
- Others are carnivores, eating other protozoa.
- Parasitic forms, involved in ecological balance, destroy other organisms.

2. *Answers may vary. The student is asked to list the positive and negative* **economic effects** *of these protozoans.*

- Protozoan diseases in man and animals cause loss through illness and death of man and animal; the expense of controlling the diseases is a serious economic drain, also.
- Saltwater forms produce limy shells that formed deposits of lime, which became limestone, used in building (e.g, the pyramids of Egypt).

➤ **Taking It Further: Importance of Protozoa**

1. *Answers may vary. The student is asked to discuss the idea of* Euglena, *or a similar organism, evolving into plants and animals and support his opinion using scientific principles.*

** Two contractile vacuoles at either end contract alternately, removing excess water and perhaps some liquid wastes.

SUPPLEMENT
Malaria and Quinine
Text pages 147–148

Malaria and Quinine (Text page 148)

➤ **Questions: Malaria and Quinine**

1. The cinchona tree became important in the fight against malaria in the 1670s; cinchona bark contains an alkaloid, called quinine, which is the effective ingredient. When quinine came into wide use, the value of the cinchona trees increased rapidly.

2. It was important to have more than one set of researchers working on the malaria problem; during WWII, when quinine was cut off by the Japanese, the U.S. government, under the direction of the Committee of Medical Research, spent a total of twenty-four million dollars on malaria research. One hundred and thirty-three universities and other organizations participated in the program. Some 1,500 workers with doctor's degrees and 4,000 laboratory technicians took part in the research. These research workers gathered facts, developed hypotheses, and tested the hypotheses by empirical means. They used scientific means to learn how to control their natural environment. Some experts believe the chief reason the U.S. won the war in the Pacific was because it controlled malaria, as well as lesser diseases, more effectively than did the Japanese.

3. A disease like malaria would flourish during a war because foreign soldiers would have to live with stressful circumstances, unsanitary conditions, and possibly open wounds in unknown places, like the jungles of Southeast Asia. Also, malaria would spread rapidly due to blood-thirsty mosquitoes.

➤ **Taking It Further: Malaria and Quinine**

1. *Answers may vary. The student is asked to research one of the many scientists mentioned in "The Control of Malaria" section in the text, and describe the role that he played in controlling malaria.*

➤ **Questions: Chapter Review**

1. In a protozoan reproducing by binary fission, there is no stage which corresponds entirely to the birth of a large animal. The last division corresponds most nearly. The cell before division might be called "parent," and the two halves, "daughters"; but, of course, the names are inappropriate.

2. Examples: *Euglena* has a long flagellum; paramecium has many cilia; amoeba sends out false feet; plasmodium has no structure designed for locomotion but is carried by the liquid in which it lives.

3. *The student has studied three examples of* **symbiosis**, *one in each of the following chapters 10, 12, and 13.*

However, he is only require to state the second two examples since the first is uncertain:

- Some fungi form a **symbiotic-like relationship** (called *mycorrhiza*) with the roots of pine trees, where the hyphae grow within the cells of the roots and extend their mycelium out into the soil; some scientists think the fungi aid the roots in absorption.
- *Lichen* is a plant structure composed of an alga and a fungus growing together in **symbiotic cooperation**; the alga (typically a green alga) provides food for the fungus (i.e., *Ascomycetes*), and the fungus provides security and moisture for the alga.
- Another example of **symbiosis** is the harmonious relationship between the *Paramecium bursaria* (a bright green paramecium) and a species of *Chlorella*; the paramecium does not have to eat because the tiny single-celled alga lives in its tissues and produces some of its food. The paramecium in turn moves about and takes the alga cells to light areas where sunlight is used to manufacture food for both. These chlorella are called *zoochlorellae*.

4. Like cells in general, a protozoan consists essentially of nucleus and cytoplasm. The latter digests and circulates the food.

5. Keeping mosquitoes away from malaria patients is effective and not so difficult as killing all mosquitoes. Malaria and yellow fever hindered and almost stopped work on the Panama Canal near the beginning of the twentieth century. Deaths from these diseases were greatly reduced, but still there are mosquitoes in the Canal Zone.

6. There is much difference between large plants and large animals; not so much contrast among tiny things. Between organisms and lifeless molecules the difference is immense!

7. A molecule's changing into a protozoan would be a great change indeed. On page 326 in the textbook there is an account of the formation of amino acid, but this substance is not a living thing.

8. Structures needed would be nucleus, food vacuoles, cytoplasm, plasma membranes, genes, contractile vacuoles, chromosomes, spindle fibers, DNA, RNA, ATP, and others. Abilities needed would be movement, digestion, assimilation, division, response to stimuli, choice of responses, and others.

9. The activity of an amoeba demonstrates that the mind of this creature moves in every direction at once, seeking food to envelop. It must be constantly on the move, directing its pseudopodia to reach out to whatever stimuli is there (light, movement, and the presence of food). This animal-like microbe is able to maintain all its life processes—locomotion, acquisition of food, digestion, elimination, respiration, and reproduction. *Additional details may be possible.*

◆ Think-Session Guide for Unit 5

Small Plants and Little Animals

Subject: Filterable virus

Purpose: To consider questions about the microbial world

a. **To the student:** While browsing through the journal *Science*, you find an article entitled "Smallest Cell Discovered." Without reading the report, you should know some things about these cells. Can you list them?

Teacher: At this point, the student should have a definition for a cell—either his own or one from the textbook. He should be led to a definition which includes the words "cell membrane."

b. **To the student:** The author of this report states that the cells are smaller than bacteria. Since he did not give details of methods, what are some ways that he could "get" this bit of data?

Teacher: Visual measurement with a light microscope; measurement with an electron microscope; filtration through membrane of known pore size.

c. **To the student:** These "cells" turn out to be non-nucleated and obligate parasites (that is, they must have a living host for survival). What other "microbes" could fit this description?

Teacher: Virus, mycoplasmas, rickettsiae, spirochetes.

d. **To the student:** If you knew that this author was a medical doctor, what other things might you be able to discover about these new "cells"?

Teacher: Is it a pathogen of human beings? Of cattle? Is there a vector? How does it affect its host? Is it a potential problem of significance to public health? Hopefully, this will raise many more questions than it answers about the "unseen world." You should be able to point out that not all microbes are "bad" but that many of them are beneficial to man.

UNIT 6
Animal Life

CHAPTER 14
Animals Without Backbones
(Text pages 151–176)

Arthropoda

Annelida

Platyhelminthes

Nematoda

Porifera

Mollusca

Coelenterata

Echinodermata

Peripatus

◆ **Suggestions for Motivation or Enrichment:**

1. Collect for display as many insects and their remains as you can find locally. Identify as many as you can by the specific names.

2. Locate a variety of sea shells, or illustrations of them, and make an oral report on the type of animal protection afforded in each case.

3. Visit a plant nursery, or examine a gardening supply catalog, and write down the kinds of safeguards recommended for the control of snails, slugs, and insects. Write out the names of the chemicals used and evaluate the dangers involved.

4. Maintain a simple aquarium with *Physa* snails and eel grass, and keep personal records of the oxygen production and also of the life habits of *Physa*.

◆ **Suggestions for Multimedia Resources:**[*]

1. *Insects—Springtails Through Wasps and Bees.* Fogware Publishing. Visit their Web site at <http://www.fogwarepublishing.com>. CD-ROM, Windows OS only; explores the significance of insects; examines the development and anatomy of insects as well as their range and diversity

2. *Introduction to Porifera.* Museum of Paleontology. Visit their Web site at <http://www.ucmp.berkeley.edu/porifera/porifera.html>. This University of California, Berkeley, Web site provides information about sponges and their life history; it has an evolutionary orientation.

3. *Explore by Animal.* Shedd Aquarium. Visit their Web site at <http://www.shedd.org/>, which provides information about various marine animals found at the aquarium; click on the "Explore by Animal" button at the top of their home page to access this information.

4. *Animal Kingdom*/VHS. Moody Videos. Visit their Web site at <http://www.moodyvideo.org>. Explore the 2,000 mile journey of the monarch butterfly from Canada to Mexico; see how the tail of a squirrel keeps him in balance during tree-top acrobatics, how the sonar of bats and dolphins enables them to navigate with remarkable precision; learn the language of the bees, inside a honeycomb; see spiders spin webs stronger than steel.

5. *University of Kentucky, Department of Entomology.* Visit their Web site <http://www.uky.edu/Ag/Entomology/enthp.htm>. The community portion of this Web site—especially the "Entomology Fact-Sheets" and "For Kids and Teachers" pages—provides helpful information about insects, with an emphasis on agricultural pests.

6. Conrad, Jim. "Animals." *Backyard Nature.* To access this Web site, visit <http://www.backyardnature.net/animals.htm>. Examines animals that are often found in typical American backyards; it has an evolutionary orientation.

◆ **Suggestions for Supplementary Reading:**[*]

1. Poirier, Jules. "The Magnificent Migrating Monarch." *Creation*, 20: 128–31. To read this article online, visit the *AiG* site at <http://www.answersingenesis.org/creation/v20/i1/monarch.asp>. The author describes how monarch butterflies navigate during their migration journeys.

2. Priest, Karl. "Ready to Prey." *Creation*, 23: 238–41. To read this article online, visit <http://www.answersingenesis.org/creation/v23/i2/prey.asp>. Discusses the physical characteristics and eating habits of the praying mantis

3. Weston, Paula. "Jellyfish: A Clever Hunter for a Creature with No Brain." *Creation*, 25: 428–31. To read this article online, visit <http://www.answersingenesis.org/creation/v25/i4/nobrain.asp>. Examines the physical characteristics, locomotion, life cycle, and eating habits of the jellyfish

4. "Migration." *The Why? Files: Science Behind the News.* To read this article online, visit <http://whyfiles.org/006migration/>. Pages 2–5 of this site provide students with information about monarch butterfly migration.

5. "Animals Index" (Access this Web page at <http://www.aqua.org/animals.html>). Visit the interac-

---

[*] The listing of these suggestions does not necessarily imply endorsement of content.

tive Web site of the National Aquarium in Baltimore, which provides information on marine animals and conservation. Select a creature from a pop-up and click "GO" for more information.

◆ Answers to Questions

Arthropoda (Text page 163)

➤ **Questions: Arthropoda**

1. The distinguishing characteristics of all arthropods are: they are joint-footed, they have exoskeletons made of chitin, their bodies are in sections called segments, their chief nerves are ventrally located, and they have open circulatory systems with a dorsally located heart.

2. The two main types of mouthparts are biting and chewing mouthparts (examples: grasshoppers and crickets) and piercing and sucking mouthparts (examples: mosquitoes and bedbugs).

3. Moths and butterflies both belong to the order Lepidoptera. They both undergo complete metamorphosis.

   They differ in that moths have larger bodies and feathery antennae, whereas butterflies have thinner bodies and smooth, knobbed antennae. While at rest, butterflies commonly hold their wings erect, but moths spread theirs flat. Also, the moths spin cocoons, but the butterflies do not.

4. *Examples of insects will vary. Only one example of each order is required.* The orders of insects mentioned above are:

   - Diptera: mosquito, housefly, midge, crane fly
   - Homoptera: cicada, leafhopper, aphid, scale insect
   - Coleoptera: ladybird beetle, weevil, other beetles
   - Hemiptera: bedbug, stink bug, chinch bug, lace bug
   - Odonata: dragonfly, damselfly
   - Lepidoptera: moth, butterfly
   - Orthoptera: grasshopper, praying mantis, katydid, cricket, roach, walkingstick, locust
   - Hymenoptera: bee, ant, wasp
   - Isoptera: termite

5. The typical appendages of the crustaceans include five pairs of legs, swimmerets, maxillipeds (such as on a crayfish) to aide the jaws, eyestalks, and antennae.

6. The main differences between arachnids and insects are that the arachnids have four pairs of legs, whereas the insects have three; the arachnids have two body parts, whereas the insects have three; and the arachnids have no antennae, whereas the insects have antennae. Also, the arachnids breathe using book lungs, whereas the insects have tracheae.

7. The differences between centipedes and millipedes are: centipedes have one pair of legs per body segment, whereas millipedes have two pairs of legs per body segment; also, centipedes have flattened bodies, whereas millipedes have cylindrical bodies; and centipedes are predators, whereas millipedes are largely scavengers.

➤ **Taking It Further: Arthropoda**

1. The distinguishing characteristics of all arthropods are: they are joint-footed, they have exoskeletons made of chitin, their bodies are in sections called segments, their chief nerves are ventrally located, and they have open circulatory systems with a dorsally located heart.

   The main descriptive difference is in the number of body parts. The insects have three body segments (head, thorax, and abdomen). The crustaceans have two body regions (cephalothorax and abdomen). The arachnids have two body segments. The Chilopoda and Diplopoda have multiple body segments (from 15 to 180). *Additional answers are possible.*

2. *The nine life processes are listed on page 143 of the textbook. Other information on characteristics of life can be found on pages 11, 12, and 122 of the textbook.* The arthropods and the nine life processes:

   - respiration: insects: spiracles and tracheae, arachnids: book lungs, crustaceans: gills
   - circulation: open circulatory system with heart pumping to the body in a blood sinus, not in blood vessels
   - response: nervous system with chief nerve on lower surface of the animal
   - nutrition: take food into mouths, digestive tracts
   - support: exoskeletons
   - reproduction: most lay eggs
   - movement: muscles
   - growth: molting
   - excretion: from digestive tract through rectum

Annelida (Text page 164)

➤ **Questions: Annelida**

1. Annelids are different from arthropods in that they have no appendages, no exoskeleton (and therefore do not molt), and a closed circulatory system (whereas an arthropod has an open circulatory system).

2. The annelids are segmented worms. They have bodies consisting of a series of recurring rings. Each segment is covered with a thin cuticle, beneath which is a layer of skin and then muscles. They have nephridia for excretion, a small brain, a ventral nerve, and a closed circulatory system.

Platyhelminthes (Text page 165)

➤ **Questions: Platyhelminthes**

1. The planarians are important to research because of their powers of regeneration.

2. The two different life cycles of a fluke are the two-host cycle and the three-host cycle. In the two-host cycle, the eggs hatch in a pond or stream,

and the larvae enter snails, where further development takes place. The larvae then leave the snails, crawl out of the water, and attach themselves to weeds or grasses. Grazing sheep then ingest them, and the worms develop to the adult stage and lay eggs. These eggs are then dropped with feces, and the cycle is repeated.

In the three-host cycle, the larva passes into a snail and from there to a fish. Fish-eating birds or animals eat the fish. The parasite reaches the adult form inside the carnivore that ate the fish.

3. The difference between a tapeworm and a fluke is that a fluke parasitizes various organs, whereas a tapeworm lives in the intestine of the host, absorbing digested material. To spread, the tapeworm breaks off in sections, and the sections pass out through the feces to be ingested by herbivores, whereas the flukes lay eggs which pass out through the feces to hatch and enter snails.

➤ **Taking It Further: Platyhelminthes**

1. Both are invertebrate worms. Annelids are segmented worms, as are the platyhelminth tapeworms. The annelid leeches are parasitic, as are the platyhelminth flukes and tapeworms.

   The annelid earthworms eat decaying vegetable matter, whereas the platyhelminth planarians eat dead animals. The annelids are tubular, whereas the platyhelminthes are flat, with no body cavity other than a stomach. The annelid leeches attach themselves to the skin and suck blood, whereas the platyhelminth flukes and tapeworms live inside the organs or digestive tracts of their hosts. The annelids have a digestive system, whereas the tapeworms do not. *Additional answers are possible.*

Nematoda (Text page 166)

➤ **Questions: Nematoda**

1. Symmetry means the correspondence in size, form, and arrangement of parts on opposite sides of a plane, line, or point. Nematodes typically exhibit bilateral symmetry, meaning the right side is like the left side.

2. It is important to thoroughly cook all meat to kill any parasites which may be present.

3. *Answers may vary.* Some of the parasites in this phylum and their hosts are: the pinworms, parasitizing human beings; *Ascaris,* a roundworm parasitizing pigs, horses, and man; hookworm, parasitizing humans; golden nematode, parasitizing potato plants; and *Trichinella,* parasitizing pigs and humans.

➤ **Taking It Further: Nematoda**

1. *Answers will vary. The student is asked to give examples of how the interrelationship of various organisms helps to maintain a healthy ecosystem.*

Porifera (Text page 167)

➤ **Questions: Porifera**

1. These organisms are only found in water because they are stationary and depend on the flow of water to bring food.

2. Sponges perform respiration by drawing water in through incurrent pores and forcing water out through excurrent pores.

3. A sponge reproduces usually by budding. Cells divide and form buds, which then break off the parent and become new sponges. When conditions are unfavorable, gemmules form, which are groups of cells surrounded by protective outer coverings. When conditions are more favorable, the gemmules develop into new sponges.

➤ **Taking It Further: Porifera**

1. *Answers will vary. The student is asked to give an example of how the interrelationship of various organisms helps to maintain a healthy ecosystem.*

Mollusca (Text page 170)

➤ **Questions: Mollusca**

1. The three main classes of phylum Mollusca are Gastropoda, Bivalvia, and Cephalopoda.

2. *Answers may vary.* Several ways in which this phylum is important to humans are for food, pearls, mother-of-pearl, and as pets (such as snails in an aquarium).

3. The shared characteristics of the three classes of molluscs discussed are that they are soft-bodied, have no jointed appendages, have a mantle and mantle cavity, and live in water or moist conditions.

4. A filter feeder propels a stream of water in and out of itself. It obtains food by drawing the water in through the incurrent siphon. The mouth is at the anterior end, with flaps on either side to conduct tiny plants and animals into the mouth; the cilia also help in directing food into the mouth.

➤ **Taking It Further: Mollusca**

1. *Answers will vary. Outside sources will be necessary. The student is asked to list the scientific pros and cons for the three different classes of molluscs having a common ancestor.*

2. Place a snail on a piece of clear glass and observe the movement of the muscular foot. Look through the thin, clear glass with a hand lens and recognize the cilia on the snail's foot. When the snail is moving, the cilia are easier to see.

Coelenterata (Text page 173)

➤ **Questions: Coelenterata**

1. The interstitial cells are so important to the continued survival of the above individuals because the

interstitial cells have the ability to grow into other kinds of cells. They are needed most to form cnidoblasts and reproductive cells, since these are lost and must be replaced.

2. The two different body types common to this phylum are the *polyp*, a tubelike form with tentacles at the top, and the *medusa*, a bell-shaped form with tentacles on the lower side of the body.

3. The organisms discussed in this chapter that experience alternation of generations include the *Aurelia* (a coelenterate) and other unnamed coelenterates. In Chapter 12, alternation of generations was seen in certain brown algae of the *Fucus* genus.

4. The appearance of nematocysts in other more highly organized animals would lend credence to the evolutionists' perspective because then the evolutionists could reason that the nematocysts developed in the "earlier" creatures and were retained by the more complex creatures, since the nematocysts were so useful for survival.

➤ **Taking It Further: Coelenterata**

1. The members of Coelenterata are animals instead of plants because they must secure their food and cannot make it themselves, as plants do. Also, all animals must have oxygen, usually securing it by means of lungs or gills. *Additional answers are possible.*

Echinodermata (Text page 175)
➤ **Questions: Echinodermata**

1. The water-vascular system is the circulatory system of the echinoderms. It consists of a circular tube with branches radiating out to the arms or to the periphery of the creature.

2. Calcareous means "consisting of calcium carbonate."

3. The distinguishing characteristics of phyla Echinodermata are that they are spiny-skinned; have water-vascular systems; have a nervous system with no brain and consisting of a nerve ring with branches extending out to the periphery; and have an internal, calcareous skeleton.

4. Embryonic similarities between mammals and echinoderms are so important to evolutionists because evolutionists seek to find relationships between more complex organisms and less complex organisms, indicating ancestry. They point to the similarity of the formation of the mouth and the formation of the body cavity in the embryonic development of the echinoderms and the mammals as indicative of an ancestral relationship.

➤ **Taking It Further: Echinodermata**

1. *Answers will vary. The student is asked to compare and contrast the way echinoderms and coelenterates*

perform the nine life processes. The nine life processes are listed on page 143 of the textbook. They are respiration, nutrition, circulation, response, support, reproduction, movement, growth, and excretion.

➤ **Questions: Chapter Review**

1. All insects in the adult stage have the characteristics of the phylum Arthropoda, of which insects comprise one class. The characteristics of this class are a body divided into head, thorax, and abdomen: also three pairs of legs in the adult. They have antennae, breathe through spiracles on the thorax and abdomen, and have an open circulatory system. All winged arthropods are insects, but not all insects have wings. Some undergo complete or incomplete metamorphosis.

2. Metamorphosis is complete when it includes four stages: egg, larva or nymph, an inactive pupa, and adult. Examples are the beetle, house fly, butterfly, and honey bee. Incomplete metamorphosis usually omits the pupa stage. Examples are the squash bug, bed bug, and grasshopper. In this type, the changes usually are gradual, and the larva resembles the adult. *Examples may vary.*

3. The appendages of an insect's head are antennae and various mouthparts; of the thorax, three pairs of legs, and wings (if any); of the abdomen, various tails and structures for laying eggs.

4. *Answers will vary.* Orders of insects which the student probably has observed are as follows: Diptera (e.g., house fly); Hymenoptera (e.g., bumble bee); Coleoptera (e.g., ladybird beetle); Lepidoptera (e.g., monarch butterfly); Hemiptera (e.g., squash bug); Odonata (e.g., dragonfly); Orthoptera (e.g., cricket).

5. Segmented worms have an alimentary canal which forms a straight tube through the body, and the flatworms have no body cavity other than a stomach.

Teacher: A segmented worm is made up of rings, the divisions between rings being marked by creases. There is much internal difference. A flatworm has a mouth at about the middle of the ventral side, a much-branched intestine, and two nerve cords. A segmented worm has a mouth at the anterior end, an intestine leading to the posterior end, and a nerve cord on the ventral side.

---

6. Since a species of nematode usually eats a certain kind of root, it is hindered by crop rotation, which replaces that kind of root with another kind.

7. Instinct is that facility of animal life that determines behavior on the basis of hereditary givens without conscious direction; it functions fully even the first time. Reason is the power to comprehend and think and to determine behavior on the basis of logical deductions.

8. *Answers will vary. The student is asked to investigate the economic importance of sponges.*

9. *Answers will vary. The student is asked to draw up a chart of illustrations of similarities of embryos of echinoderms and vertebrates.* In both groups, the mouth is not formed from the blastopore, but is formed separately later. Formation of the body cavity is from pouches off the earliest cavity of the embryo. The names of the stages of the embryo development are the same, although there are significant differences in the formation processes. Similarities do not prove relationships.

Teacher: A good special assignment for some students is to copy the early stages of the embryo of a starfish; for other students, the early stages of a pig or rabbit. (Stick drawings are not too hard.) The differences are considerable, especially in the formation of the gastrula stage.

---

10. The characteristics that determine the phylum Coelenterata are radial symmetry, two layers of cells, tentacles, and nematocysts. All are aquatic.

Supplement (Text page 176)

➤ **Questions: Supplement**

1. This unusual creature has been classified as a mollusc, as an annelid worm, and as an arthropod. Some have also classified it as a phylum all its own, Onychophera.

2. The jaws of arthropods could be considered modified appendages because they have parts that protrude.

3. Characteristics of the *Peripatus* that are unlike either arthropod or annelid are that its skin has a velvety texture that sheds water, and its legs are fleshy and not jointed.

CHAPTER 15
Animals With Backbones
(Text pages 179-192)

Mammals

Birds

Reptiles

Amphibians

Fishes

◆ **Suggestions for Motivation or Enrichment:**

1. Inquire of a plant nurseryman or landscape gardener about local rodent problems. Prepare a chart of available countermeasures, noting safety, cost, and dependability.

2. Trace over a world map the migratory patterns of such animals as salmon, Arctic tern, and eels (from both European and American waters). Display your map.

3. Share in a discussion of some personal experiences with snakes. Demonstrate the first aid procedures found in the most recent American Red Cross text.

4. Ascertain from the local Audubon Society or Sierra Club which forms of animal wildlife are in most danger of extinction and what practical measures are advised. Plan a workable strategy.

◆ **Suggestions for Multimedia Resources:**[*]

1. Conrad, Jim. "Animals," *Backyard Nature*. To access this Web site, visit <http://www.backyardnature.net/animals.htm>. Examines animals that are often found in typical American backyards; it has an evolutionary orientation.

2. *Animals*. Shedd Aquarium. Visit their Web site at <http://www.shedd.org/>. Provides information about various marine animals found at the Shedd Aquarium; click on the "animals" button at the top of their home page to access this information.

3. *Animal Kingdom*/VHS. Moody Videos. Visit their Web site at <http://www.moodyvideo.org>. Explore the 2,000 mile journey of the monarch butterfly from Canada to Mexico; see how the tail of a squirrel keeps him in balance during tree-top acrobatics, how the sonar of bats and dolphins enables them to navigate with remarkable precision; learn the language of the bees, inside a honeycomb; see spiders spin webs stronger than steel.

4. *Introduction to Vertebrates*. Fogware Publishing. Visit their Web site at <http://www.fogwarepublishing.com>. CD-ROM; Windows OS only; examines fishes, amphibians, reptiles, birds, and mammals; evolution

5. *Birds: Cranes Through Passerines*. Fogware Publishing. Visit their Web site at <http://www.fogware-publishing.com>. CD-ROM; Windows OS only; Discover why cranes are noted for their elaborate courtship displays and how the California gull became the state bird of Utah; learn about puffins, doves, parrots, owls, thrushes, hummingbirds, woodpeckers, orioles, and many more.

6. "Checklist of Amphibian Species and Identification Guide." *Northern Prairie Wildlife Research Center*. Access this site at <http://www.npwrc.usgs.gov/resource/herps/amphibid/index.htm>. Provides valuable information on reptiles for both teachers and students; helpful field guide gives both text and pictures of fauna and flora of southern California; it has an evolutionary orientation.

7. Buchheim, Jason. "A Quick Course in Ichthyology." *Odyssey Expeditions*. To access this course

---

[*] The listing of these suggestions does not necessarily imply endorsement of content.

online, visit <http://www.marinebiology.org/fish.htm#how%20fish%20swim>. Provides information about fish, including how they swim, eat, breathe, and sense in their environment; some pictures; it has an evolutionary orientation.

8. "Reptiles." *San Diego Natural History Museum.* To access this site, visit <http://www.sdnhm.org/exhibits/reptiles/index.html>. Provides information on reptiles for both teachers and students; helpful field guide gives both text and pictures of fauna and flora of southern California; evolution

9. Lewis, Deane. *The Owl Page.* To access this Web site, visit <http://www.owlpages.com/>. Provides a comprehensive introduction to owls around the world, including photographs and bird calls

10. *Digital Frog 2.* Digital Frog International. Visit their Web site at <http://www.digitalfrog.com/products/index.html>. CD-ROM; Windows and Macintosh OS. This interactive CD is a frog dissection, anatomy, and ecology program. A workbook is available as part of an educational package. The company has a special home school price.

◆ Suggestions for Supplementary Reading:[*]

1. Weston, Paula, and Carl Wieland. "Bears Across the World." *Creation,* 20: 4, 28–31. To read this article online, visit <http://www.answersingenesis.org/creation/v20/i4/bears.asp>.

2. Hennigan, Tom. "Snakes Alive." *Creation,* 20: 4, 40–42. To read this article online, visit <http://www.answersingenesis.org/creation/v20/i4/snakes.asp>.

3. Weston, Paula. "Bats: Sophistication in Miniature." *Creation,* 21: 1, 28–31. To read this article online, visit <http://www.answersingenesis.org/creation/v21/i1/bats.asp>.

4. Weston, Paula. "Turtles." *Creation,* 21: 2, 28–31. To read this article online, visit <http://www.answersingenesis.org/creation/v21/i2/turtles.asp>. The author shows how turtles resist evolutionary explanation.

5. Sarfati, Jonathan. "The Non-evolution of the Horse." *Creation,* 21: 3, 28–31. To read this article online, visit <http://www.answersingenesis.org/creation/v21/i3/horse.asp>.

6. Weston, Paula. "Spectacular, Surprising Seals." *Creation,* 22: 4, 28–32. To read this article online, visit <http://www.answersingenesis.org/creation/v22/i4/seals.asp>.

7. Weston, Paula. "Sharks: Denizens of the Deep." *Creation,* 23: 2, 46–50. To read this article online, visit <http://www.answersingenesis.org/creation/v23/i2/sharks.asp>.

8. Catchpoole, David. "Wings on the Wind." *Creation,* 23: 4, 16–23. To read this article online, visit <http://www.answersingenesis.org/creation/v23/i4/migration.asp>.

9. Sarfati, Jonathan. "A Coat of Many Colours: Captivating Chameleons." *Creation,* 26: 4, 28–33. To read this article online, visit <http://www.answersingenesis.org/creation/v26/i4/chameleon.asp>.

10. "Migration." *The Why? Files: Science Behind the News.* To read this article online, visit <http://whyfiles.org/006migration/>. Pages 7–10 of this site provide students with information about bird migration.

◆ Answers to Questions

Mammals (Text page 182)

➢ **Questions: Mammals**

1. Oviparous means that the animal lays eggs. The duck-billed platypus is oviparous.

   Viviparous means that before birth the young are attached to the mother by a placenta, receiving their nourishment from the bloodstream of the mother. *Most mammals are viviparous so examples will vary.* Examples would include dogs, bears, and monkeys.

   Ovoviviparous means that the young are nourished inside the mother's body by a small yolk sac (no placenta) and then born alive. The baby then crawls into a pouch and drinks milk until it is more fully developed. Opossums and kangaroos are ovoviviparous.

2. The distinguishing characteristics of mammals are that they have hair, milk glands, and highly differentiated teeth. Also, nearly all of them bring forth their young alive, and all are warm-blooded, or endothermic.

3. Monotremes would have been difficult to classify because they lay eggs as birds and reptiles do but also suckle their young as mammals do.

4. The rumen is a large storage compartment of the stomach. In the rumen, the food is mixed with bacteria that produce enzymes that can break down cellulose. This is a first step in digestion for ruminants.

➢ **Taking It Further: Mammals**

1. *Answers will vary. The student is asked to write a short opinion essay, discussing the idea that man is the "deadliest predator" with respect to the Creation Mandate.*

Birds (Text page 184)

➢ **Questions: Birds**

1. Birds are oviparous, meaning that they lay eggs.

2. The characteristics of birds that make them lighter than other animals of comparable size are the air

---

[*] The listing of these suggestions does not necessarily imply endorsement of content.

sacs in their bodies, hollow bones, and hollow feathers.

3. Down feathers have a loose, soft structure and are for insulation.

   Contour feathers are responsible for covering and protecting the body and giving the bird its color.

   Flight feathers are contour feathers that extend past the body of the bird. Their shafts are hollow with veins called barbs that are interlocked and hooked together.

➤ **Taking It Further: Birds**

1. A bird has compact organs, a large heart that beats rapidly, and strong muscles which work intensely to achieve flight. To keep up its energy, the bird must eat concentrated food frequently and in large quantities.

Reptiles (Text page 188)

➤ **Questions: Reptiles**

1. Some general characteristics of the class Reptilia are that they have scales, claws, and eggs. They are also cold-blooded, or ectothermic.

2. Snakes and lizards differ by the fact that lizards have legs and snakes do not.

3. It is important to remain still and calm if bitten by a snake because the venom is spread through the blood, and faster circulation is therefore dangerous. Also, hemotoxin affects the ability of the blood to transport oxygen, and therefore the individual must not exert himself.

4. Alligators have broad, U-shaped heads and overlapping upper jaws, which hide the teeth in their lower jaws. Crocodiles have narrow, more pointed snouts, and their upper and lower jaws have interlocking teeth which are always exposed. Crocodiles are also more aggressive than alligators.

➤ **Taking It Further: Reptiles**

1. Birds and reptiles are both vertebrates that lay eggs. Most reptiles have claws, as do birds. However, birds have feathers and most can fly, whereas reptiles have scales, and no living reptile can fly. Most birds incubate their eggs, whereas reptiles leave their eggs to be incubated by the temperature of the environment. Birds are endothermic (warm-blooded), whereas reptiles are ectothermic (cold-blooded). *Additional answers are possible.*

2. *The student is asked to classify the snapping turtle (kingdom to species).*

Teacher: An outside source is necessary to answer this problem because there is no appendix at the back of the textbook and the appendix at the back of this manual does not provide the entire classification of the snapping turtle. An excellent online resource is the

Animal Diversity Web; visit this site at <http://animaldiversity.ummz.umich.edu/site/index.html>.

The classification of a snapping turtle is as follows:

Kingdom: Animalia

Phylum: Chordata

Subphylum: Vertebrata

Class: Reptilia

Order: Testudines (or Chelonia)

Family: Chelydridae

Genus: *Celydra*

Species: *serpentina*

Amphibians (Text page 190)

➤ **Questions: Amphibians**

1. The respiratory system of the amphibians is special in that the skin is used in respiration. Some newts respire entirely through the skin, and frogs increase their oxygen intake this way.

2. The tongue of the frog is a complex of muscle fibers of great agility. It is remarkable in that it can be extended, wrapped about a fly, and retracted into the mouth so quickly that it is hard to see how the catch was made.

3. It would be a problem if deoxygenated and oxygenated blood mixed in the heart of an amphibian because the deoxygenated blood needs to go to the lungs and skin, but the oxygenated blood needs to go to the rest of the body. If they mixed, the body would not receive the oxygen it needs, and the animal would be very weak from lack of oxygen.

➤ **Taking It Further: Amphibians**

1. Some evolutionists would consider *Amphibia* to be a transition species because some amphibians live part of their lives in water and part of their lives on land. This may indicate to the evolutionists that amphibians represent a transition stage in which aquatic creatures became adapted to land living.

Fishes (Text page 192)

➤ **Questions: Fishes**

1. The three classes of fish discussed above are Chondrichthyes, Osteichthyes, and Agnatha.

2. The differences between sharks and bony fish are:

   • Sharks have a skeleton of cartilage, without any bone, whereas bony fish have skeletons made of true bone.
   • Sharks have no operculum covering the gills, as bony fish have.
   • The young of many sharks are born alive, whereas the bony fish lay eggs which hatch.
   • Sharks' scales are of a different type from those of bony fish, and are very rough.

- Sharks do not have swim bladders, so they sink when not swimming. Bony fish have swim bladders which help them float.

3. *Answers may vary.* Several ways in which fish are important to man are for food, as a source of protein and vitamins; as a major industry; as a pattern for streamlining (as in shapes of ships or missiles); and as pets. Fish are important in another sense, in that destructive fish (such as the sea lamprey) need to be controlled, and dangerous fish (such as the shark) need to be avoided by the average person.

➤ **Taking It Further: Fishes**

1. Amphibians and fishes are both vertebrates that do not maintain a constant temperature. Most amphibians and fishes lay eggs. All fishes live in the water and have gills, and some amphibians do for at least part of their lives. Amphibians are moisture-loving creatures. Some fish and some amphibians eat insects.

   However, most fishes have scales, and amphibians do not. Fish have fins, and amphibians have limbs. Fish live in the water all their lives, and amphibians live only part of their lives in water. *Additional answers are possible.*

➤ **Questions: Chapter Review**

1. Some characteristics of mammals are that they have hair; the females have milk glands to nourish their young; and nearly all mammals bring forth their young alive, an exception being the platypus.

2. In addition to the mammals, other animals, including some fishes, lizards, snakes, and insects give birth to living young, as alluded to on page 179 of the text, but these species are not named. Sharks are specifically named on page 191.

3. A ruminant can eat grass quickly in a dangerous area or when passing through good pasturage, then regurgitate and chew it again in a safer or more barren area.

4. *Answers may vary.* Some reasons why man has domesticated various mammals include: some, such as cows, sheep, and pigs, are good for food; some, such as dogs and cats, are good for pets; some, such as sheep, provide wool or fur; and some, such as horses and oxen, do work.

   There may be several reasons why especially mammals have been domesticated. They have enough intelligence to relish the care which is given them. They learn to give up the resentment which an animal has at being fenced in. The likeness to persons suggests the food and care which should be given to these captives.

5. Down feathers have a loose, soft structure and are for insulation.

   Contour feathers are responsible for covering and protecting the body and giving the bird its color.

Flight feathers are contour feathers that extend past the body of the bird. Their shafts are hollow with veins called barbs that are interlocked and hooked together.

6. Before the bird molts, pin feathers begin to grow. Blood flows through them as they develop. As the young feathers grow longer, the blood supply collects only in the bases of the shafts. The tips of the shafts encase the feathers with a thin, waxy sheath. When full grown, the feathers become lifeless. Then, as it molts, the bird preens them to remove the sheaths.

7. If it should be proved that the ancestors of snakes had legs, this would hinder the theory of an upward tendency in evolution because it would be an example of an animal becoming less complex, or moving in a downward direction, by losing limbs.

8. Certain amphibians need gills when they are larvae because they live in the water as larvae and therefore need to breathe through gills.

9. A frog takes air into the mouth cavity through the nostrils, then forces it down into the lungs by lifting the floor of the mouth cavity. The nostrils, mouth, and esophagus are kept closed while forcing the air down. There is no mechanism drawing the air into the lungs. If the mouth were open, the air would just pass out through the mouth because there would be no force on the air.

10. A frog has a tongue attached at the anterior end, leg muscles enabling it to jump long distances, skin fitted for respiration, a distinctive call, and a form in its immature stage that is different from that of the adult.

11. Mammals and birds, being warm-blooded animals, keep almost the same temperature all the time. This production of heat uses a big portion of the food. A toad, on the other hand, remains at the temperature of its surroundings and can get along with but little food when it is inactive.

12. If the frog were to become extinct, the animals on which it preys might become more numerous. Examples are flies, mosquitoes, worms, beetles, and bugs.

13. A large lake without plant life would not be a good habitat for fish because the fish need oxygen and food, and the plants give off oxygen and provide homes and food for insects, which are in turn food for the fish.

14. Eels from both North America and Europe go to a certain place in the Atlantic Ocean to lay their eggs. Young eels migrate to the land of their parents and swim up the streams. The streams flowing into the Atlantic are easier to find than the ones

flowing into the Gulf of Mexico. The journey is shorter and safer, also.

15. Many fish eggs fail to be fertilized by sperm and even more are eaten by other animals. Young fish are devoured in large numbers. Fish have prodigality of reproduction; without this prodigality, they would become extinct.

16. In a creek there is much shallow water, and fish congregate in the deep pools or among boulders. Both depth and obstructions afford some protection. In a lake, water plants give protection and serve as food for little animals on which the fish feed.

◆ Think-Session Guide for Unit 6

Animal Life

Subject: Freshwater creatures

Purpose: To develop interest in unfamiliar animals

a. **To the student:** A zoologist made a collection of animals from a freshwater stream. He could easily identify all but three of the specimens. If you were the collector, what steps would you take to identify them?

Teacher: 1) Look for similarities with known organisms.
2) Consult with another scientist.

---

b. **To the student:** Laboratory study indicated that two of the unknown animals used gills for oxygen while the third used an air tube. What can you say about the relationships now?

Teacher: The two are probably closer related to each other than to the third.

---

c. **To the student:** The two-gilled creatures "got sick and died" during the first week, whereas the air breather developed into a mosquito. What may have accounted for the deaths?

Teacher: Lack of food, lack of oxygen, disease, unfavorable environment, old age.

---

d. **To the student:** What kind of information is still missing for the gilled creatures?

Teacher: Does it have legs? Is it an immature or mature form? Could it be a hybrid of a previously undescribed form?

At this point some numbers may be in order. Did you realize there are 700,000 species of insects? 25,000 species of Crustacea? 8,000 species of Annelida (segmented worms)? 20,000 species of "bony fishes"? 2,800 species of amphibia?—just to give some possibilities.

UNIT 7
The Biology of Man

CHAPTER 16
Form and Major Functions of the Human Body
Text pages 195-211

Skeletal System
Muscular System
Circulatory System
Respiratory System
Digestive System
Excretory System
Integumentary System

◆ Suggestions for Motivation or Enrichment:

1. Inquire about the requirements for athletic training for the Physical Education department of a local university or community college. Prepare to talk to your class or home school group as if you were considering them to be potential athletes applying for school sports.

2. Keep a calorie record of your actual daily food intake and record your average count beside your height and sex on a class chart. Put in parentheses the calorie stipulation given for you on a standard diet or health chart.

3. On prominent hand or arm veins, press distally along a vein until you can see that the valves are stopping return flow. Sketch the vein-valve pattern and prepare to describe the value of such valves.

4. Study the process of mouth-to-mouth artificial respiration, as from a current American Red Cross brochure, and give a demonstration. (Do not actually breathe into a person's mouth when there is no emergency.)

◆ Suggestions for Multimedia Resources:[*]

1. *Human Life*/Video (Moody Science Classics, 62 min.). Traces how each person develops into a wonderful masterpiece

2. *Red River of Life*/Video (Moody Science Classics, 30 min.). Examines the circulatory system using microscopic photography

3. *My Body, Myself.* (Fogware Publishing. Visit <http://www.fogwarepublishing.com>). CD-ROM; Windows OS only. Presents a virtual museum with interactive exhibits and information on the health, structure, and processes of the human body

4. *Human Anatomy Online.* (Inner Learning Online. Visit <http://www.innerbody.com/index.html>). This site explains the working of the various systems of the human body to students, using a mix of text and illustrative graphics.

◆ Answers to Questions

Skeletal System (Text page 197)
➤ Questions: Skeletal System

1. The axial skeleton includes the skull, the vertebral column, the ribs, and the sternum. The appendicular skeleton includes the shoulders, hips, and limbs.

2. The different types of joints named in the text and where they are found in the body are as follows:
   - Pivot joint: neck
   - Ball-and-socket: where the arms and legs attach to the pectoral and pelvic girdles
   - Hinge joints: elbows and knees
   - Gliding joints: wrists and vertebrae
   - Sutures: between the skull bones

3. The patella is a bone which is part of the knee. It is enclosed in a ligament and increases leverage in the knee. It also protects the vulnerable knee joint.

4. The main functions of the skeletal system are: providing support and shape for the body, protecting vital organs, providing levers for the muscles to move the body, and producing new red blood cells.

➤ Taking It Further: Skeletal System

1. *Outside sources will be necessary. The student is asked to identify the types of joints, other than those mentioned, that can be found in the body.*

2. A forensic anthropologist can help the police positively determine the identity of a crime victim from his skeletal remains because the skeletal remains can provide information such as gender, age, race, and stature. Often teeth can also help to positively identify the individual.

Muscular System (Text page 200)
➤ Questions: Muscular System

1. An antagonistic pair of muscles is a pair of muscles that work against each other. When one contracts, the other must relax, and vice versa.

2. *Examples may vary.* The different types of muscle tissues are as follows:

---

[*] The listing of these suggestions does not necessarily imply endorsement of content.

- Skeletal muscles attach to the skeleton and are the primary muscles used to produce motion. They have alternating light and dark fibers and are called striated muscles. Muscles in the arms and legs would be examples.
- Smooth muscles occur in sheets in the walls of internal organs. They have no striations. Muscles in the intestinal tract would be an example.
- Cardiac muscles are found in the heart. They are a highly branched network of muscle fibrils.

3. The body gets the energy to perform muscle contractions from oxidation of food. However, under greater exertion, when oxygen cannot reach the muscles fast enough, they perform lactic acid fermentation.

4. ATP, adenosine triphosphate (containing three groups of phosphorus), stores energy. The removal of one phosphorus group from ATP releases energy to be used by the muscle. ADP, adenosine diphosphate (containing two groups of phosphorus), is left. ADP has less energy than ATP and needs to be built up again into ATP, which serves as an agent of power.

5. The Sliding Filament Theory accounts for muscle contractions by saying that ATP activates a reaction between the two filaments of striated muscles, producing actomyosin. When this happens, the two filaments slide alongside each other, causing the muscle to contract.

6. The main functions of the muscular system are to produce movement, to help maintain posture, to change the shapes of organs, and to produce heat.

➤ **Taking It Further: Muscular System**

1. The skeletal system is vital to the function of the muscular system because the skeletal system gives shape and support to the body. Muscles attach to the bones and move the bones. *Additional detail is possible.*

Circulatory System (Text page 203)

➤ **Questions: Circulatory System**

1. The main organ of the circulatory system is the heart.

2. Arteries carry blood from the heart, while veins return blood to the heart. They are connected by the capillaries.

3. The different components of blood are the plasma (liquid) and the corpuscles (cells).

Plasma is water containing nutrients, wastes from cells, hormones, proteins, and several other substances.

The red corpuscles have hemoglobin, a compound of oxygen, hydrogen, carbon, nitrogen, and iron. The oxygen is carried to the cells of the body, where it is used for oxidation of food.

White corpuscles are less numerous than the red and devour foreign particles or produce antibodies that chemically destroy foreign particles.

Platelets function in forming clots to prevent blood loss from cuts.

4. It is important to know a patient's blood type because antigen A must be kept away from antibody a, and antigen B from antibody b. Otherwise, the defense mechanism of the antibodies will be triggered. The antibodies in the blood cause the clumping of red corpuscles, and the result is that they cannot pass through the capillaries. *Examples will vary.*

5. The functions of the lymphatic system are to carry food and oxygen to the cells, to destroy dead blood cells, to filter the blood, and to produce white corpuscles and antibodies that fortify the body's immunity.

6. The main functions of the circulatory system are to transport oxygen and food to the cells; to remove excess water, carbon dioxide, and other wastes from the cells; to regulate body temperature; and to maintain the body by preventing infection.

➤ **Taking It Further: Circulatory System**

1. The circulatory and skeletal systems are interdependent because the skeletal system provides shape and support for the body and also produces red blood cells, while the circulatory system spreads nutrients throughout the body, including to the bones. *Additional details are possible.*

Respiratory System (Text page 204)

➤ **Questions: Respiratory System**

1. The components of the respiratory system are the nostrils, sinuses, trachea, epiglottis, bronchi, lungs, and diaphragm.

2. The muscles involved in respiration are unique in that they have two control mechanisms. Usually a person's breathing is involuntary, although he can also consciously exert control. He can choose to breathe deeply or hold his breath, but he can hold his breath only to a point before the involuntary mechanism takes over. The involuntary mechanism maintains breathing while sleeping. Since continued breathing is vital to life, this involuntary characteristic is vital.

3. The functions of the respiratory system are to take in oxygen and to expel excess carbon dioxide, heat, and water.

➤ **Taking It Further: Respiratory System**

1. It is harder to breathe at higher elevations because the air has less concentration of oxygen. Therefore, a normal breath will not bring in the same amount of oxygen as at lower elevations where

there is a higher concentration of oxygen. *Additional details will vary.*

2. It is important for people to be good stewards of plant life because plants give off oxygen, which humans and animals need to breathe. Plants also provide food for humans and animals, whether directly or indirectly. *Additional details will vary.*

Digestive System (Text page 208)

➤ Questions: Digestive System

1. Smooth muscle tissue aids in most of the digestive process by contracting in a wavelike motion, called peristalsis, and pushing the food through the alimentary canal.

2. Digestion begins in the mouth, where the teeth break down the large pieces of food and saliva from the three salivary glands near the mouth begins the breakdown of starches. Once the food is chewed well, the tongue pushes it back to the throat, where it is swallowed.

    The food passes down through the esophagus (by peristalsis) and into the stomach. The stomach continues the digestion of food by producing digestive enzymes and then twisting to mix the enzymes with the food. When the food reaches the consistency of a thick paste, called chyme, the pyloric valve opens and allows it to pass to the small intestine.

    The majority of digestion occurs in the first one third of the small intestine using enzymes from the intestinal linings, gall bladder, liver, and pancreas. In the remaining two thirds of the small intestine, absorption of the digested food into the bloodstream occurs.

    In the large intestine, liquid is absorbed, and the concentrated wastes are passed into the rectum for storage. At a convenient time, the rectum muscles contract, forcing the feces out through the anus.

3. The function of the liver in digestion is to provide bile to the small intestine for the digestion of fats.

4. The vitamins mentioned in this chapter that are necessary to human diets and what they are thought to influence are as follows:

    *   Vitamin A: vision, secreting powers of mucous membranes
    *   Vitamin B_1 (thiamine): nerves
    *   Vitamin B_3 (niacin): skin, nerves, muscles, mental health
    *   Vitamin B_6 (pyridoxine): metabolism
    *   Biotin: skin, digestion, nerves
    *   Vitamin B_{12}: formation of red corpuscles
    *   Vitamin C: intercellular substances such as connective tissue fibers, intercellular cement, matrix of bone dentine
    *   Vitamin K: blood coagulation
    *   Vitamin D: bone growth

5. The main function of the digestive system is to break down and convert food to substances that can be used by the body for energy, repair, and growth.

➤ Taking it Further: Digestive System

1. Of the systems studied so far in the textbook, the following systems are necessary for the proper functioning of the digestive system:

    *   The blood of the circulatory system receives the digested food and transports the nutrients to the cells of the body.
    *   The smooth muscles of the muscular system provide the peristalsis action which moves the food along the alimentary canal.
    *   The skeletal muscles also move the jaw of the skeletal system to achieve chewing action.
    *   The respiratory system takes in the oxygen needed to burn for energy to achieve any action in the body, including digestion.
    *   The impulses from the nervous system stimulate the muscles involved in digestion.

    Additional answers are possible.

Excretory System (Text page 209)

➤ Questions: Excretory System

1. The organs used by the excretory system are the kidneys, urinary bladder, and large intestine. The sweat glands are also used.

2. *The student is asked to draw and label the different sections of a nephron. See textbook page 209.*

3. The benefit of eliminating only highly concentrated urine is the conservation of water which is needed by the body.

4. The purposes of the excretory system are the removal of harmful waste products and the conservation or excretion, as conditions demand, of such normal components as water, sugar, and salts.

➤ Taking It Further: Excretory System

1. If the excretory system ceased to function, the wastes would not be removed, throwing off the balance of the blood composition. There would be an excess of nitrogen and other harmful substances, which would kill the cells of the body. Also, there would be an excess of normal substances such as water, sugar, and salts. The body would not live long after the excretory system failed.

Integumentary System (Text page 211)

➤ Questions: Integumentary System

1. The main organ of the integumentary system is the skin.

2. The sebaceous gland is the main source of acne problems.

3. It is important for people with little melanin to wear suntan lotion on a cloudy day because the UV rays from the sun penetrate through the clouds, and a person with little melanin is more vulnerable to the harmful effects of the UV rays.

4. The main functions of the integumentary system are to protect the body, to control body temperature, to function as an excretory organ, and to prevent the body from drying out.

➤ **Taking It Further: Integumentary System**

1. *Outside sources may be used to answer this question.*

➤ **Questions: Chapter Review**

1. Tendons usually are found at either end of a muscle, attached to bones, enabling the muscle to move one of the bones by shortening itself. A ligament is similar in structure to a tendon, and it reaches from bone to bone.

2. The stimulation of smooth muscles is different from that of striated muscles in that striated or skeletal muscles are under *voluntary control* and smooth muscles are under *involuntary control* and function automatically without conscious effort

---

Teacher: A striated or skeletal muscle usually is controlled by one nerve. An impulse over this nerve causes the muscles to contract, and when the impulse ceases, the muscle relaxes. A smooth muscle is provided with two nerves. Impulses over one nerve cause contraction; over the other, relaxation. This information is taught in the next chapter. See page 214.

---

3. Flexing a limb, such as bending at the elbow or knee, is accomplished as follows: A muscle is attached at one end to a bone for anchorage. At the other end, the muscle has a tendon that passes over a joint and attaches to another bone. When the muscle contracts (shortens), the second bone moves, flexing the limb.

4. Skeletal muscles generally work in antagonistic pairs, each one counteracting the other. To hold the arm in one position, pairs of muscles are in a state of partial contraction; that is, only part of the muscle fibers are in a state of contraction in each muscle.

5. From the left ventricle, the blood goes into the aorta, to the arteries, to the capillaries of the head, to the veins, to the right atrium (auricle), to the right ventricle, to the artery that goes to the lungs, to the capillaries of the lungs, to the veins that come from the lungs, to the left atrium, finishing the circuit by entering the left ventricle.

   Blood goes also from the left ventricle to the aorta, to the arteries, to the capillaries of the foot, to the veins, to the right atrium, to the right ventricle, finishing the circulation in the same way as the blood from the head, going to the lungs and back.

6. Pulse is the rushing of blood due to the contraction of a ventricle. The contractions force blood into the arteries, whereas in the veins the blood is under much less pressure.

7. If the blood flows from a wound in spurts, a good sized artery has been severed.

8. Arms do not hang downward so much of the time as legs do; therefore, there is not so much danger of blood failing to return to the heart and going the wrong way.

9. The protective devices found in the respiratory system are as follows:

   • Near the opening of the nostrils there are hairs, which filter the dust from the air.
   • The nasal passages and sinuses located further back warm and humidify the air before it reaches the lungs.
   • Lining the inside of the trachea is a mucus secretion, which collects dust and foreign particles. Cilia, embedded in this lining, sweep these particles upward to the throat, where they are swallowed.
   • The cough reflex may expel or dislodge large particles, such as food.
   • Normally in swallowing food the epiglottis covers the top of the larynx, preventing food from going down the windpipe.
   • The lungs are enclosed in a moist pleural membrane, which protects them by reducing friction on their surface.

10. The main reasons why the body needs food are for growth, repair, and energy for movement and body processes. *Discussion should include examples.*

11. Membranes, which line the intestines, keep undigested food from entering the bloodstream. When the food has been made soluble by digestion, it can pass through such membranes.

12. The skin, lungs, kidneys, and large intestine function in excreting waste products.

13. The filtrate that passes into the tubules contains many substances that are necessary for the body and are reabsorbed, such as water, sugar, and salts. The capillary bed surrounding the tubule is the site of reabsorption.

CHAPTER 17
Body Controls and Human Reproduction
Text pages 213-232

Nervous System

Endocrine System

Homeostasis

Reproductive System

Narcotics, Alcohol, and Tobacco

Behavior

◆ Suggestions for Motivation or Enrichment:

1. Discover the optical correctives represented by your family and the types of defects involved. Contrast lens varieties and note any legislation pertaining to lens construction in your state or in other states.

2. Plot your personal temperature record on a graph, taking your readings at regular intervals during your waking hours for one week (avoid periods immediately after eating or drinking). Try to account for variations. Explain the implications of "normal temperature."

3. Display and discuss any material available on drugs or alcohol from local police or other agencies.

4. Obtain a chicken trachea within minutes after slaughter. Carefully scissor it open from bottom to top and pin it open on a board, tilted to raise the trachea's upper end. Apply India ink, in 75% normal saline, to the bottom edge of the trachea, and watch the rising color as the cilia continue beating upward. Ask a smoker to blow fresh cigarette smoke through a narrow tube on one side only of the trachea, above the color line. Notice the effect on the color rising. Apply the experimental results to the smoker's personal well-being.

◆ Suggestions for Multimedia Resources:[*]

1. *Human Life*/Video (Moody Science Classics, 62 min.). Traces how each person develops into a wonderful masterpiece

2. *My Body, Myself.* Fogware Publishing. Visit <http://www.fogwarepublishing.com>. CD-ROM; Windows OS only. Presents a virtual museum with interactive exhibits and information on the health, structure, and processes of the human body

3. *Human Anatomy Online.* Inner Learning Online. To access this Web site, visit <http://www.inner-body.com/index.html>. This site explains the working of the various systems of the human body to students, using a mix of text and illustrative graphics.

---

[*] The listing of these suggestions does not necessarily imply endorsement of content.

4. Chudler, Eric H. *Neuroscience for Kids.* To access this University of Washington Web site, visit <http://faculty.washington.edu/chudler/neurok.html>. Provides a wide variety of information and activities about the nervous system

◆ Suggestions for Supplementary Reading:[*]

1. Savige, Craig. "Electrical Design in the Human Body." *Creation,* 22: 143–45. To read this article online, visit <http://www.answersingenesis.org/creation/v22/i1/electrical.asp>.

2. Farley, Dixie. "Bone Builders: Support Your Bones with Healthy Habits." *Kid Source Online.* To read this article online, visit <http://www.kid-source.com/kidsource/content4/bone.build-ers.fda.html#Eat>. Valuable information about good health and nutrition for young people

3. Batten, Donald. "Red-blooded Evidence." *Creation,* 19: 224–25. To read this article online, visit <http://www.answersingenesis.org/creation/v19/i2/blood.asp>. The author refutes the evolutionary "seawater" argument.

4. Calkins, Joseph. "Design in the Human Eye." *Bible Science Newsletter,* March 1986. To read this article online, visit the *Creation Moments, Inc.* site at <http://www.creationmoments.com/articles/article.php?a=59>.

◆ Answers to Questions

Nervous System (Text page 217)

➢ **Questions: Nervous System**

1. The three divisions of the nervous system and their main functions are as follows:

   - The central nervous system consists of the brain and spinal cord. The brain is the control center for the body. The spinal cord is the main nerve cluster and also governs reflexes.
   - The peripheral nervous system includes the sense organs and the vast network of nerves, which carry messages to and from the brain and spinal cord.
   - The autonomic system consists of nerves leading to the internal organs and their ganglia scattered among these organs. These nerves control the action of the internal organs.

2. The different sense organs of the body are eyes for sight, ears for hearing and balance, tongue and nose for taste and smell, and skin for touch.

3. Rods respond to dim light, and cones are sensitive to colors and to objects in bright light.

4. The Eustachian tube connects the middle ear to the throat, and thus air pressure on both sides of the eardrum is equalized.

➢ **Taking It Further: Nervous System**

1. *Outside sources will be necessary. The student is asked to research and list the names and functions of the twelve cranial nerves.*

Endocrine System (Text page 220)

➤ **Questions: Endocrine System**

1. Hormones are the secretions of the endocrine glands. They are sometimes referred to as "chemical messengers" since they carry instructions for the various body functions.

2. Feedback systems are so important for endocrine glands because the glands need an indication as to how much of their hormones to produce. The amount produced needs to vary based on conditions in the body.

3. *See page 219 of the textbook for a list of the glands, the hormones they produce, and their main functions.*

4. A deficiency of the hormone insulin is the cause of certain types of diabetes.

5. *Answers may vary.* Several life processes that are controlled or regulated by the endocrine system are growth and maturation, metabolism, excretion, and blood pressure.

➤ **Taking It Further: Endocrine System**

1. Both the nervous and endocrine systems regulate body function in general and organ function in particular. Both respond to stimuli.

   However, the nervous system receives stimuli from the environment and can produce an immediate response to them. If the stimulus stops, the response stops. Prolonged controls over the body functions are produced by the endocrine glands. The endocrine glands secrete hormones which are spread through the circulatory system, whereas the nervous system has its own system of specialized nerve cells to transmit messages. *Additional answers may be possible.*

Homeostasis (Text page 224)

➤ **Questions: Homeostasis**

1. Homeostasis means "remaining the same." It is applied to the internal conditions of living things. It refers to the balance of body fluids, temperature, chemical composition, etc. maintained by the body. Without this balance, an organism would cease to function properly, and the organism would die.

2. A feedback mechanism is a system which responds to a certain stimulus and adjusts to the conditions present. An example would be in the regulation of temperature. A certain temperature is to be maintained, and if the temperature drops or rises, the body responds to correct the imbalance. *Examples may vary.*

3. Hunger is based on the need for food. Appetite is the desire for food (often regardless of need).

4. Cybernetics is a term applied to the science of controls in both living and nonliving systems. Homeostasis applies to living systems only.

Reproductive System (Text page 227)

➤ **Questions: Reproductive System**

1. When fertilization of eggs and development of embryos take place outside an animal, there is much loss of life due to predation and environmental conditions. Internal fertilization and development protect the developing embryo.

2. The testes, prostate, ovaries, and uterus are the organs involved in reproduction.

3. *See the chart on page 226 of the textbook for the menstruation cycle.*

   The FSH from the pituitary gland starts the growth and maturation of the egg. The growing follicle produces estrogen, which builds up the endometrium of the uterus. Ovulation (release of the egg) takes place at the middle of the cycle. If the egg is fertilized, the corpus lutem is formed, which produces progesterone, which maintains the endometrium for the fertilized egg. If the egg is not fertilized, the endometrium is sloughed off, and this is called menstruation.

➤ **Taking It Further: Reproductive System**

1. The endocrine system regulates the maturation of the reproductive organs. It also regulates by means of hormones the functioning of the reproductive organs. For example, the pituitary gland releases the FSH, which causes the growth and maturation of the egg in the female. *Examples may vary.*

➤ **Questions: Chapter Review**

1. The nervous system of an insect is much simpler than that of a human. In an insect, the nerve cord is ventral, or on the underside, whereas in humans it is on the dorsal, or back, side. The insect nervous system consists of several clumps of nerve cells, called ganglia, which are found along the nerve cord; and the largest one, found in the head, is called the brain. Humans also have masses of nerve cells called ganglia. The nerves also branch out throughout the body. The human brain is much more complicated than that of the insect, however. *Additional detail is possible.*

2. If a person's hand accidentally touches a hot stove, for instance, heat receptors in his hand send an impulse along a sensory neuron that has its cell body in a ganglion near the spinal cord. Another fiber of this sensory neuron extends into the spinal cord, where it forms a synapse with a connector neuron. This second neuron makes another synapse with the end of the proper motor neuron, which has a fiber reaching down into his arm that causes the muscle to contract and lift his hand

from the stove. Reflex arcs protect the body, since the distance that the impulse travels is less than the distance to the brain, thereby shortening the amount of time before one responds. *See also the diagram on page 214 of the textbook.*

3. The functional unit of the nervous system, the neuron, consists of a cell body and long, very slender fibers extending from it. The cell body is gray in color. The fibers, however, and the outer portion of the spinal cord, have white coverings. Thus, the outer part of the brain and the inner part of the spinal cord are called gray matter, and the nerves and fibers are called white matter.

4. Some materialists attempt to classify all human acts as reflex. If they are not single reflex acts, they are combinations; if an act is not an immediate response to a stimulus, it is a delayed reflex. Such reasoning rules out voluntary acts and makes a person wholly dependent upon his environment. This is not proved.

5. Sound waves in the air strike the ear drum, which is the outer wall of the middle ear. A chain of three tiny bones (called the anvil, the hammer, and the stirrup) carry the sound waves across to the inner ear, where the liquid in the cochlea is caused to vibrate. Inside the cochlea is a membrane that vibrates at a particular place for each frequency. When this membrane vibrates, it stimulates hairlike structures, which then initiate a nerve impulse that goes to the brain, where the sound is interpreted.

6. Likeness of form is one indication of kinship or descent from a common ancestor. Creationists have said that God, in His creation, repeated certain types of organs. The "kinship," then, between man and chicken is in their having a common Creator.

7. The nervous system regulates and coordinates body activities through nervous impulses conducted by the nervous system. The endocrine system regulates and correlates body activities through hormones carried by the blood and lymph. Both systems operate to enable an organism to maintain a constant internal environment.

8. The pituitary gland is called the master gland because of its control over several other endocrine glands.

9. Endocrine glands secrete their hormones directly into the bloodstream. They do not have ducts or tubes. Digestive glands secrete by way of ducts into the alimentary canal.

10. Reproduction is the only life process that does not directly contribute to the well-being of the individual organism that performs that function.

SUPPLEMENT A
Narcotics, Alcohol, and Tobacco
Pages 228–229

Narcotics, Alcohol, and Tobacco (Text page 229)
➤ **Questions: Narcotics, Alcohol, and Tobacco**

1. A stimulant speeds the action of the heart or of some other organ of the body. A narcotic (depressant) dulls the action of some part or all of the body. Adrenaline is a natural stimulant, and alcohol is a narcotic. *Examples may vary.*

2. Some health risks in consuming alcohol are inflammation of the liver, delirium tremens (hallucinations), psychiatric problems, and accidents or injuries due to loss of muscle coordination and lack of judgment.

3. Nicotine is the substance found in cigarettes that is addictive. Tar is the substance in cigarettes that causes cancer.

SUPPLEMENT B
Behavior
Pages 229–232

Behavior (Text page 232)
➤ **Questions: Behavior**

1. Behavior is the external response of the organism to environment. Behavior is what the organism does.

2. A tropism is the response of a sessile (stationary) organism, such as a plant. A taxis is the response of a motile organism, such as a paramecium. A negative tropism or taxis is a turning away from a stimulus. A positive tropism or taxis is a turning toward a stimulus.

3. Innate behaviors are reflexes and instinct (including feeding, reproductive behavior, migration, hibernation, and estivation).

4. Instinct is unlearned behavior, whereas the conditioned reflex is a learned type of behavior. The conditioned reflex is a reflex in which the stimulus is different from the inborn one but produces the same result.

♦ **Think-Session Guide for Unit 7**

The Biology of Man

Subject: *Muscles and bones*

Purpose: *To introduce some thoughts about a human body system*

a. **To the student:** In man, muscles that move the forearm are located in the upper arm. Of what significance is this fact?

Teacher: This question should lead to an explanation of the fact that flexor and extensor types of muscles invariably work across a joint. Given the parts (muscles, bones, and a joint), this design is necessary to the action of the whole. Engineering and leverage laws support the fact that the parts must be arranged as they are to do their job.

b. **To the student:** With your palm up and without bending your wrist, make a tight fist; then open your hand flat. Repeat. Where are the muscles that move your fingers into a fist?

Teacher: The main muscles for finger movement are located in the forearm. Some small muscles of the hand are also involved, but they are secondary and can be ignored here.

c. **To the student:** Of what significance is the fact that the muscles which control the movement of the hand are in the forearm?

Teacher: The point here is: If the whole mass of muscles moving the fingers were located in the hand itself, the hand would have to be several times larger to have the same strength. The hand would therefore be clumsy. Is the ape hand different anatomically from man's? To extend the point, because finger muscles are located in the forearm, there is a greater dexterity without a great loss of strength.

UNIT 8
Plant Life

CHAPTER 18
Plants Without Conducting Systems
(Text pages 235-239)

Bryophyta

Mosses

Liverworts and Hornworts

◆ Suggestions for Motivation or Enrichment:

1. Research and describe reports in news magazines of the Moscow peat fires of July and August 1972. Discuss peat as fuel.

2. From a commercial peat moss sack, note the source of such a supply and the indications of value for gardening. Present your findings in your class or home school group as if you were a gardening consultant.

3. Be ready to discuss the popular use of the word "moss" with respect to color. Obtain sample cards from a hardware store or home improvement center. Notice the various shades of moss-green.

4. From a moss layer on a rock or from an excellent picture in the textbook (page 234) extrapolate as to the probable effects of moss growth.

◆ Suggestions for Multimedia Resources:*

1. Conrad, Jim. "Backyard Plants." *Backyard Nature.* To access this Web site, visit <http://www.backyardnature.net/botany.htm>. This is a menu of links to pages on plants often found in backyards; this site has an evolutionary orientation.

◆ Suggestions for Supplementary Reading:*

1. Kent, Livija. "Bryophytes." *Connecticut River Homepage* Web site. To access this site, visit <http://www.bio.umass.edu/biology/conn.river/bryophyt.html>. It includes links on liverworts and mosses, with text and illustrations for both.

◆ Answers to Questions

Mosses (Text page 237)

➤ **Questions: Mosses**

1. List and describe the various structures of moss.

   - **Gametophyte plant:** has leaves and bears eggs and sperm
   - **Sporophyte plant:** bears spores and has no leaves but is simply a stalk with a spore case growing from

the apex of the gametophyte plant; a mature sporophyte is made up of a foot, stalk, and capsule

   - **Apical:** tip cell that produces new cells
   - **Antheridia:** male gametophytes for mosses that have separate male and female gametophytes; forms sperm that are later released in stalked sacs
   - **Archegonia:** female gametophytes for mosses that have separate male and female gametophytes; bears the eggs of mosses; vessel-shaped, on stalks
   - **Calyptra:** a portion of an old archegonium carried as a cap atop a growing sporophyte
   - **Operculum:** The sporophyte capsule that contains moss spores; the capsule falls off with the calyptra when the spores are ripe
   - **Rhizoids:** rootlike projections that are sent into the soil

2. *Alternation of generations* refers to organisms that alternate between sexual and asexual reproduction cycles. The gametophyte generation is dominant among mosses. The gametophyte plant has leaves and bears eggs and sperm, while the sporophyte plant bears spores and has no leaves, having only a stalk with a spore case that grows from the apex of the gametophyte plant.

3. Rhizoids differ from roots by not conducting water.

4. Ways in which water is important to the survival of mosses are as follows:

   - The fertilization of an egg by a sperm requires the presence of water because the sperm must swim a considerable distance to reach the egg.
   - A few mosses live in water.
   - Spores need moist conditions to germinate.

➤ **Taking It Further: Mosses**

1. *Answers may vary. The student is asked to explain why moss might be used to help fill a lifeless landscape.*

2. *Answers may vary. The student is asked to explain why mosses do not grow to be very tall plants.*

Liverworts and Hornworts (Text page 239)

➤ **Questions: Liverworts and Hornworts**

1. **Similarities**

   - Liverworts and mosses are both bryophytes.
   - Most liverworts and mosses are small, green, terrestrial plants that thrive in moist areas.
   - Some liverworts have a leaflike and stem-like organization in their gametophytes, as do mosses.
   - Liverworts and mosses both have a distinct alternation of two generations.

   Differences

   - Liverwort stems lie flat against the substrate while most moss stems are erect.
   - The sporophyte of most liverworts is not as large or complex in development as that of a moss.

- In liverworts, the spore grows directly into the gametophyte body without going through the threadlike protonemal stage as in the moss life cycle.

2. The hornworts' sporophyte generation is unique because it has an erect, rodlike sporophyte that contains a core of mother cells that produce spore quartets by meiotic division, enabling spores to be continually formed for a long time. In addition, the sporophyte can live independently of the gametophyte because it is photosynthetic.

➤ **Taking It Further: Liverworts and Hornworts**

1. *Answers may vary. The student is asked to explain if it is possible to make a case for liverworts and hornworts being more primitive than mosses.*

➤ **Questions: Chapter Review**

1. A spore of a moss is formed in the capsule at the top of the sporophyte. It is much like a mold spore in that if it grows it will become a plant. A sperm is produced on the gametophyte or leafy plant and will not produce a plant unless it unites with an egg. Then it grows and becomes a sporophyte.

2. Both mycelium and protonema are networks, but the protonema is green, whereas the mycelium has no chlorophyll. From the protonema grow leafy plants, while the growths from the mycelium are much like itself.

3. The gametophyte has rootlike rhizoids and green leaves and produces eggs and sperm. The sporophyte is a stalk and capsule perched on a gametophyte plant.

4. *Marchantia* thrives in shade and moist atmosphere. It may be found at the edge of a swamp or under the projecting rocks of a cliff.

5. A spore produces a protonema which may spread and be divided. In addition, each plant growing from a protonema may produce buds or leaf portions which may become new plants.

6. Mosses and lichens may grow with scant or no soil and may escape death in dry places or in dry periods by becoming inactive. A lichen is a combination of fungus and alga, whereas a moss is a bryophyte.

CHAPTER 19
Plants With Conducting Systems: Structure and Growth
(Text pages 241-258)

Plants with Tubes: Tracheophyta
Cells, Leaves, and Photosynthesis
Roots and Absorptions
Stems and Translocation
Plant Growth Substances

◆ **Suggestions for Motivation or Enrichment:**

1. From a local area, collect and label various types of leaves. Note the fragility or hardiness in relation to the climatic or seasonal conditions.

2. In a local produce market, list in columns the displayed foods which are leaves or stems or root-parts.

3. Find a cross section of a tree branch or trunk that shows tree rings of growth. Discuss the implications which the rings seem to suggest.

4. Prepare a word analysis of scientific words, as you divide some words and note the separate meanings (e.g., photo/synthesis, inter/node, trans/location, herbi/cide, epi/dermis, gameto/phyte).

◆ **Suggestions for Multimedia Resources:**[*]

1. Conrad, Jim. "Backyard Plants." *Backyard Nature.* To access this Web site, visit <http://www.backyardnature.net/botany.htm>. This is a menu of links to pages on plants often found in backyards; this site has an evolutionary orientation.

2. *What is Photosynthesis?* ASU Photosynthesis Center Web site. Visit <http://porphy.la.asu.edu/photosyn/education/learn.html>. It provides a basic introduction to photosynthesis for young people.

3. *Gallery of Illinois Plants.* To access this Illinois Natural History Survey Web site, visit <http://www.inhs.uiuc.edu/cwe/illinois_plants/>. This site provides pictures of numerous plants found in Illinois and a list of botany Web sites.

4. *Plants Database.* Natural Resources Conservation Service, United States Department of Agriculture. Visit <http://plants.usda.gov/index.html>. This site is a database of plants, providing substantial information about plants in the United States.

◆ **Suggestions for Supplementary Reading:**[*]

1. Oard, Michael J. "A Uniformitarian Paleo-environmental Dilemma at Clarkia, Idaho, USA" *Answers in Genesis,* April 2002. Visit <http://www.answersingenesis.org/tj/v16/i1/clarkia.asp> to read this article online. It discusses one of the most remarkably-preserved plant fossil localities in the world.

2. Williams, Alexander. "Kingdom of the Plants: Defying Evolution." *Answers in Genesis,* December 2001. To read this article online, visit <http://www.answersingenesis.org/creation/v24/i1/plants.asp>.

3. "Basics of Tree ID" *Web site of the Virginia Tech College of Natural Resources, Department of Forestry.* To search this site, visit <http://www.cnr.vt.edu/dendro/forsite/Idtree.htm>. This site is a basic guide to identifying trees; it describes leaves, fruits, barks, with many images.

---

[*]   The listing of these suggestions does not necessarily imply endorsement of content.

4. Douma, Michael, ed. "Why Are Plants Green?" *Institute for Dynamic Educational Advancement*. Visit <http://webexhibits.org/causesofcolor/7.html> to read this article online. The editor and contributors write about color in the plant world.

5. Conrad, Jim. "Tree Bark" *Backyard Plants*. To read this article online, visit <http://www.backyardnature.net/treebark.htm>.

6. "Why Do Leaves Change Color in the Fall?" *Science Made Simple*. To read this article online, visit <http://www.sciencemadesimple.com/leaves.html>. This article explains how leaves change colors and prepare for winter. It includes science projects.

◆ **Answers to Questions**

Plants with Tubes: Tracheophyta (Text page 243)

➤ **Questions: Plants with Tubes: Tracheophyta**

1. Ferns and roses have several differences. Roses are flowering plants and ferns are not. Ferns resemble bryophytes in their method of reproduction, emphasizing sporophyte generation. Roses, on the other hand, utilize sexual reproduction.

Teacher:	The answer found above provides an answer based on a general description of flowering plants, assuming that roses are representative of flowering plants in general. It is based on the sections of chapter 19 covered by this set of questions and is all that is required of the student.

2. The tubes found in the stems of plants that conduct water from region to region are known as *vascular tissue*. A fine pathway of *xylem* cells, equipped to carry water and mineral salts to leaves or other organs, is formed inside each growing stem. A food supply line known as *phloem* also forms inside each young shoot organ and enables foods formed by the plant to flow from one plant organ to another.

   Leaves form along the stem and capture the sun's energy, forming sugar molecules. This is accomplished by ventilating pores, which allow gases to move within the sandwiched layers of cells inside; when sunlight falls on green leaf cells, carbon dioxide and water are changed to produce energy-rich sugar molecules by *photosynthesis*.

Teacher:	The answer found above is based on the sections of chapter 19 covered by this set of questions and is all that is required of the student. However, he should also read sections 19.6, 19.13, 19.14, and 19.15 to be able to better answer the question.

3. The flower of a plant is similar to the animal reproductive system.

4. *The answer found below is based on the sections of chapter 19 covered by this set of questions and is all*

that is required of the student. However, the student should read sections 19.9 and 19.14 to be able to better answer the question about meristematic tissue.

Meristematic tissues are the growing zones of cells in plants. The rootcap is a small cover that protects the probing root growth from being cut by sand grains in its path.

➤ **Taking It Further: Plants with Tubes: Tracheophyta**

1. Water travels from the roots to the leaves because the moisture in the soil is exposed to the surface area of the roots. The molecules of moisture are passed into the roots and through the roots to the rest of the plant through the xylem.

   Food produced by the plant, on the other hand, travels from leaves to roots. The food is produced in the leaves and flows from plant organ to plant organ along the phloem.

Cells, Leaves, and Photosynthesis (Text page 247)

➤ **Questions: Cells, Leaves, and Photosynthesis**

1. Differences between animal and plant cells are as follows:
   - Plant cells usually form rigid walls, while animal cells generally have soft walls.
   - Plant cells may form large and distinct central vacuoles, which are not normally encountered in animal cells.
   - Plant cells usually contain unique bodies such as the chloroplasts in which photosynthesis takes place.
   - In the division of plant cells, the cytoplasm is separated into two daughter cells by a division plate that partitions the old cell into two chambers from inside outward. Animal cells, on the other hand, are "pinched apart" by a deepening cleavage furrow from the outside inward.
   - Meiosis in plant life cycles ordinarily takes place in spore formation, whereas animal meiosis usually occurs in the formation of sperms and eggs.

2. Various types of leaf margins and venation are as follows:
   - **Margin:** smooth, rounded, toothed, lobed, crisped, or combinations of these border designs
   - **Venation:** netting or branching, parallel

3. Main functions of guard cells are as follows:
   - Guard cells possess many chloroplasts that serve as cellular centers of starch storage.
   - Guard cells function as leaf air valves.
   - The process of photosynthesis begins with the guard cells.

4. The main function of leaves is to capture the sun's energy to form energy-rich sugar molecules.

5. The steps of photosynthesis are as follows:
   - Some of the light falling on the plant enters the chloroplasts of green cells.

- Chlorophyll molecules in the chloroplasts are then able to utilize this light and use its energy in important chemical changes.
- Light energy is then used in a process of splitting water molecules, known as a photolysis reaction, which results in the release of oxygen.
- Some of the released oxygen is diffused out of the cell, while the remaining oxygen may be used in cellular respiration or other cell processes.
- The hydrogen released in the splitting of water becomes attached to a molecule known as a hydrogen carrier.
- At this time in the chloroplast, carbon dioxide molecules become attached to molecules known as carbon carriers.
- With the hydrogen and chemical energy supplied by photolysis, the carbon carrier changes from a sugar chain that is five carbon atoms long and contains phosphorous into two sugar molecules consisting of three carbon atoms each. The sugar molecules are attracted to phosphorous and are called sugar phosphate molecules.

➤ **Taking It Further: Cells, Leaves, and Photosynthesis**

1. Plants might be considered more advanced organisms than animals because they make their own food.

Roots and Absorptions (Text page 251)

➤ **Questions: Roots and Absorptions**

1. The two main types of root systems are the *taproot system* and the *fibrous root system.*

2. The different structures and sections of a typical root are as follows:
   - **Zone of cell division:** made up of the rootcap and the meristem, cells in this zone are left behind as cell division continues to produce new cells
   - **Zone of cell enlargement:** cells that are left behind begin to grow and change their shape by becoming either part of the epidermis layer or cortex tissue
   - **Zone of root hairs:** root hairs and other structures begin to grow; root hair develops from a bulge on the outside of an epidermal cell that extends like a finger into the soil mass
   - **The differentiating root:** contains the conducting region known as the stele and two surrounding cylindrical layers of cells known as the pericycle and the endodermis; a few cells in the pericycle are capable of differentiation or the growth of a new branch root; cork cambium forms within the pericycle in the mature root
   - **Xylem:** found within the pericycle column; lies at the center of the root; carries water within the root
   - **Phloem:** found within the pericycle column; enfolded between the grooves in the xylem region; carries sugars, amino acids, and some other substances
   - **Zone of secondary tissues:** also known as the maturation zone

3. A large surface area is important for root function because it aids in the absorption of materials from the soil. Most absorption occurs in the younger terminal regions near the root tips, especially along the root hairs.

4. Meristematic tissue is found at the root tip and in the vascular cambium.

Teacher: The answer found above is based on the sections of chapter 19 covered by this set of questions and is all that is required of the student. However, the student should read section 19.14 to be able to better answer the question about meristematic tissue.

---

5. The definitions and differences between *diffusion* and *osmosis* are as follows:
   - *Diffusion* is the net movement of molecules from regions of their greater concentration to regions of their lesser concentration.
   - *Osmosis* is the diffusion of water molecules across a selectively permeable membrane.

   Osmosis operates on the basis of the diffusion of water. However, the selectively permeable barrier does not allow for the free diffusion of all substances; only water can cross the barrier to diffuse.

➤ **Taking It Further: Roots and Absorptions**

1. *Answers may vary. A short essay comparing root structure to the foundation of a tall building is requested.*

Stems and Translocation (Text page 256)

➤ **Questions: Stems and Translocation**

1. The internal structure of a typical stem is as follows:
   - The outer layer of a stem is an area of cortex tissue sometimes known as bark.
   - The cortex tissue surrounds a ring of vascular cylinders.
   - At the center of the stem is the pith, a core made of thin-walled cells.
   - At the tip of each stem is the apical meristem or meristematic region that provides new clusters of cells that grow into the tissues of an adult stem.

2. Primary tissues are cells within each vascular cylinder that differentiate into an outer zone of phloem tissue, an inner area of xylem tissue, and a ribbon of cambium tissue between the phloem and xylem areas.

   Secondary tissues are found in the stem of a perennial. As the stem of a perennial grows, the cambium within divides, forming new phloem cells outside and new xylem cells inside. In addition, cambium tissue forms from cells so that a whole cambium cylinder exists, producing new cells in a ring. New cells formed outside become the secondary phloem, and those inside become secondary xylem.

3. The various functions of a stem are as follows:
   - A stem provides support for the plant organs.
   - The stem acts as a conductive organ for water, mineral ions, and food to pass.

- The green stems of herbs carry out photosynthesis and act as food-supplying organs.
- The tissues of the stem often function as areas of food storage.
- Some stems have a special shape or structure that equips them to play a specific role in the life of a particular plant.

4. Translocation is the movement of water and food up and down the plant. Without translocation, the plant will die.

➤ **Taking It Further: Stems and Translocation**

1. *Answers may vary. The student will not actually be expected to carry out such an experiment.*

Plant Growth Substances (Text page 258)

➤ **Questions: Plant Growth Substances**

1. Dr. Went's experiments proved that plant growth was stimulated by a substance found in the tips of leaves and stems. He discovered the growth subsantance IAA (Indole-3-acetic acid).

2. The functions of plant growth substances are as follows:

   - **Indole-3-acetic acid (IAA):** This is one of the most important natural growth substances found in plants. IAA is produced by cells near the coleoptile tip and moves down the leaf, increasing the growth of cells directly beneath the tip.
   - **2,4-Dichlorophenoxyacetic acid (2,4-D):** This is an artificial growth substance that can be used as an herbicide if applied in high concentration.

➤ **Taking It Further: Plant Growth Substances**

1. *Answers will vary. The student is asked if the comparison of plant growth substances to animal hormones is a good one; he also is to explain why or why not.*

➤ **Questions: Chapter Review**

1. The cells in a root are surrounded by a membrane which allows water to pass through by diffusion. There are salts inside the cells which are not allowed to pass through the membrane to the outside. While this condition exists, more water passes into the cells, causing pressure which makes the water rise.

2. By diffusion, salts would enter roots through the membranes until the concentration of salts in the root is the same as that in the soil water outside. But, by active absorption, the root can take in much more salt than the degree of concentration in the soil water.

3. IAA is necessary for the development of cells. Light coming from one side alone drives IAA to the opposite side of the stem, causing the stem to turn toward the light due to the additional growth on the dark side.

4. During the growth period in summer, a new layer of bark or phloem and a new layer of wood or xylem are formed.

5. Palisade cells usually are shown in a single cross section of a leaf (Figure 19-11). The cells look like a palisade of tall posts around a fort. We forget about the solid array of cells behind the ones we see.

6. A stoma is the place between two long, slender cells called guard cells in the epidermis of a leaf. The stoma is closed when the guard cells straighten and lie side by side.

7. To recognize the type of veining, look at the small veins. Usually plants which come up with two cotyledons have the small veins of the leaf arranged in a network. Those that come up with one cotyledon have the small veins running parallel to each other.

CHAPTER 20

Plants With Conducting Systems: Flowers, Seeds, and Fruits

Text pages 261–288

Purpose of Flowers, Seeds, and Fruits

Flowers

Seeds

Fruits

Vascular Plants Without Seeds

Seed Plants

◆ **Suggestions for Motivation or Enrichment:**

1. Discuss allergies that are represented in your family, class, or home school group; and discover the apparent connections to various pollens. Locate the offending plants on a local map, marking the seasonal variations each person reports.

2. Consult a wildlife guidebook or other reference book, and list the wild plants that are considered safe for survival diet or for exotic supplement.

3. Collect weed seeds by walking through a dense, dry, weeded area while wearing rough, exposed socks. Analyze the seed structures that you have gathered and study them with a hand lens. Sketch the various dispersal devices.

4. Compare and contrast some real flowers with some artificial ones of the same name. Do the same with some real flower fragrance and some commercial perfume or talcum powder of the same naming.

◆ **Suggestions for Multimedia Resources:**[*]

1. Conrad, Jim. "Backyard Plants." *Backyard Nature.* To access this Web site, visit <http://www.backyardnature.net/botany.htm>. A menu of links to

---

[*] The listing of these suggestions does not necessarily imply endorsement of content.

various pages on plants often found in backyards; this site has an evolutionary orientation.

2. *Plants and Animals, Partners in Pollination.* To access this Smithsonian Education Web site, visit <http://smithsonianeducation.org/educators/lesson_plans/partners_in_pollination/index.html>. This site was developed to help educators teach a lesson on the relationship between plants and animals in pollination and its importance to mankind.

◆ **Suggestions for Supplementary Reading:***

1. Chapman, Geoff. "Orchids … a witness to the Creator." *Answers in Genesis.* To read this article online, visit <http://www.answersingenesis.org/creation/v19/i1/orchids.asp>.

2. Armstrong, Wayne P. "Blowing In The Wind: Seeds & Fruits Dispersed By Wind." *Wayne's Word.* To read this article online, visit <http://waynesword.palomar.edu/plfeb99.htm>.

3. "What Makes Popcorn Pop?" *National Aeronautics and Space Administration.* To read this article online, visit <http://www.nasa.gov/audience/forkids/home/popcorn.html>.

4. Amato, Ivan. "Save the Flowers: Would-be Scent Engineers Aim to Resurrect Lost Floral Fragrance." *Science News Online*, September 2005. To read this article online, visit <http://www.sciencenews.org/articles/20050924/bob10.asp>. The author discusses the waning fragrance of roses and other ornamental-flower varieties.

◆ **Answers to Questions**

Flowers (Text page 269)

➤ **Questions: Flowers**

1. Determinate inflorescence refers to plants in which the flower stem terminates its growth when the first flower forms. Indeterminate inflorescence refers to plants in which the flower stem continues to grow after one flower forms, and many new flowers are produced above it.

2. Photoperiods are periods of light, usually referring to the length of light in a day. They are important because a slight change in photoperiods can trigger or suppress the production of flowers.

3. The typical flower has four parts: the sepals, the petals, the stamens, and the pistil.

   • **Sepals:** The sepals are the outermost flower organs. They are usually green in color and look like leaves. The sepals protect the inner flower organs when the flower is still rolled up in the bud.
   • **Petals:** The petals are brightly colored and are found inside the sepals. The color of the petals aids in sexual reproduction by attracting insects to the flower,

as does the nectar produced by the petals of some flowers.

   • **Stamens:** The stamens contribute directly to the reproduction. Fully developed stamens are typically shaped like miniature croquet mallets and consist of a filament attached to the flower stalk below and an anther attached to the top of the filament. Spore-forming cells called sporogenous cells are found within the anther. The spores turn into pollen grains and are released by the opening of the anther.
   • **Pistils:** One or more pistils are found at the top or center of a flower. Along with the stamens, pistils are also directly involved in reproduction. Each pistil consists of three main parts—the stigma, the style, and the ovary.

   The stigma is a small cellular surface at the top of the pistil. Sweet, sugary fluid secreted by hair cells and gland cells on the stigma aids in the capture and growth of pollen grains. The style is the tubular portion that extends from the stigma to the ovary. The ovary is an expanded cavity containing ovules that will develop into seeds.

4. If a plant produces perfect flowers, it would be classified as a monoecious plant.

5. The differences between monocots and dicots are as follows:

   • **Leaf veins:** Monocots have parallel veins, while dicots have netted veins.
   • **Flower:** Monocots usually have flower parts in threes or multiples of three, while dicots usually have flower parts in fours, fives, or multiples of fours or fives.
   • **Vascular bundles:** Monocots have vascular bundles that are scattered, while dicots have vascular bundles that are in a peripheral ring.
   • **Seed leaves:** Monocots have one seed leaf, while dicots have two seed leaves.

   Examples will vary. The **monocots** include the grasses, palms, lilies, irises, sedges, and others. The **dicots** include the rose, elm, oak, maple, sunflower, and many other flowering plants.

6. The processes of self-pollination, cross-pollination, and double fertilization are as follows:

   • **Self-pollination:** Self-pollination occurs when pollen of one flower is deposited in the stigma of the same flower or the stigma of another flower on the same plant.
   • **Cross-pollination:** Cross-pollination occurs when pollen of one flower is deposited in the stigma of a flower of another plant.
   • **Double fertilization:** Double fertilization involves two sperms. One sperm unites with the egg nucleus of the embryo sac to form a diploid zygote. The other sperm unites with the two polar nuclei in the center of the embryo sac to form one endosperm nucleus.

➤ **Taking It Further: Flowers**

1. Wind-carried pollen would be of most importance in causing hay fever.

* The listing of these suggestions does not necessarily imply endorsement of content.

2. Flowers that benefit from long photoperiods would do well in the summer in Alaska and Siberia.

Seeds (Text page 272)

➤ **Questions: Seeds**

1. The structures of a typical seed are the seed coat, endosperm tissue, radical, plumule, and cotyledon(s).

2. Ways seeds are disseminated are as follows:
   • Some seeds travel through the air by means of hairy pappus, tufts, blades, or wings.
   • Some seeds travel either on the inside or outside of an animal or man.
   • Some seeds float on water.
   • Some seeds are hurled into the air by their plants with explosive force.

3. The seeds of some plants are dormant until the conditions are right for growth. This prevents seeds from germinating until conditions for seedling growth are adequate.

   The seeds of many desert plants remain dormant due to inhibitors that prevent germination. It typically takes a half inch of rain or more to wash off the inhibitors to enable the seeds to germinate, ensuring that the plants have sufficient moisture to survive.

   Other seeds may be dormant until their seed coats are weakened, the seeds are stored at low temperature for a sufficient period of time, or they receive sufficient light once the soil has been disturbed. In some cases, such dormancy is necessary to enable the embryo within the seed to mature.

➤ **Taking It Further: Seeds**

1. *Answers may vary.*

Fruits (Text page 275)

➤ **Questions: Fruits**

1. A fruit is a ripened ovary, along with any other flower parts that develop with it. A simple fruit forms from one flower that has only one pistil. An aggregate fruit develops from one flower with several pistils that ripen together. A multiple fruit arises from many flowers that grow together and form one fused fruit.

2. Pollen grains and pollen tubes supply IAA to the immature ovary, thus preventing abscission. This enables the ovary to develop into fruit.

3. Seedless fruits are usually produced by artificially adding growth substances to flowers that are not pollinated.

➤ **Taking It Further: Fruits**

1. *Answers will vary. A short essay is requested.*

➤ **Questions: Chapter Review**

1. There are several different types of indeterminate flower clusters. They include the *raceme*, the *spike*, *corymb*, the *umbel*, and the *head*.

   • **Raceme:** the basic indeterminate inflorescence; each flower has its own stalk or pedicel; younger flowers grow above older flowers
   • **Spike:** similar to the raceme except that its flowers do not have their own pedicel; each flower on a spike is sessile
   • **Corymb:** similar to the raceme except that the flower stalks have different lengths, giving the cluster a flat or sloping top
   • **Umbel:** similar to the corymb, but the internode distances between each pedicel are extremely short
   • **Head:** similar to the umbel except that the flowers are sessile or nearly so

2. A pollen grain develops from a cell in the stamen, then reduces its chromosomes to half the number contained in the other plant cells. When placed on a stigma of a flower, it forms a long tube which grows to an ovule in an ovary. Three nuclei are involved; the tube nucleus stays in the tube while the two sperm nuclei enter the ovule. One sperm nucleus unites with the egg nucleus, forming the embryo plant, while the other sperm nucleus unites with the polar nuclei to form the stored food, called endosperm.

3. Seeds have a protective covering, stored food, and an embryo plant which already has developed to a great extent. A mold spore is simply a cell which has great powers of growth. It is easily seen that the spores are very vulnerable.

4. A seed needs warmth, moisture, and oxygen to start its growth. Soil is not needed until later. Seeds may be placed under different environments to demonstrate these needs.

5. The knowledge of plant reproduction is very helpful to man because it helps man produce food crops and grow plants for our enjoyment.

SUPPLEMENT
Comparative Study of Fossil and Living Vascular Plants
Pages 276–288

Vascular Plants Without Seeds (Text page 284)

➤ **Questions: Vascular Plants Without Seeds**

1. It is hard to determine the age of rock formations because no one was there at the time to establish its date of formation.

2. *Psilotum* and *Rhynia* plants are both vascular plants without seeds, have grasslike structures, spores, and central xylem zones. The main reasons for not grouping them together are as follows:

   • *Psilotum* have definite leaves and roots, but the *Rhynia* do not.
   • *Psilotum* stems have a star-shaped protostele similar to flowering plants.

- Along the sides of some *Psilotum* stems are intricate three-parted sporangia that differ markedly from the simple spore sacs of *Rhynia* fossils.
- *Psilotum* is more like the lycopods or the horsetails.

3. Ferns have been used as food, decorative plants, and medicine.

➤ **Taking It Further: Vascular Plants Without Seeds**

1. *Answers will vary. A short essay is requested.*

Seed Plants (Text page 288)

➤ **Questions: Seed Plants**

1. A polyphyletic origin refers to organisms that are unrelated to each other by descent.

2. A living fossil is a living organism that looks like a specimen found in the fossil record.

3. The *telome theory* asserts that branches and leaves evolved from short stems known as telomes. However, the presence of ginkgo leaf fossils in the same rock strata as plants with simple telomes undermines the possibility that leaves and stems evolved from telomes.

4. A conifer is a cone-bearer. Its *life cycle* is as follows:

- The pine sporophyte is a large, coniferous evergreen with slender, elongated leaves, which arise in clusters on short branches.
- Each year, cones are formed on other short branches. The larger female, or ovulate cone, has husky scales in which are two winged ovules.
- In the center of each ovule is a megaspore mother cell that divides by meiosis, yielding four monoploid megaspores. Only one of these four spores functions, whereas the others disintegrate. The megaspore divides many times by mitosis, forming a minute female gametophyte.
- Cones with smaller scales form pollen grains that are carried by the wind, some of which fall in between the scales of the young ovulate pinecones.
- The pollen grains land near the ovule on a drop of sticky, resinous material. As this sticky material dries, it contracts, pulling some pollen grains back into the tiny hole at the end of the ovule.
- Pollination occurs during the first year the female cones have been formed. Over a long period of time, the pollen grain forms a pollen tube that absorbs nutrients from the cells of the ovule.
- Inside the pollen tube (male gametophyte), two sperm nuclei form. The pollen tube containing sperm nuclei grows toward the egg inside the female gametophyte.
- As a pollen tube enters an egg cell, one sperm nucleus unites with the egg nucleus in fertilization, which occurs about one year after pollination (the cone is already over a year old).
- During the growing season following fertilization, a pine embryo begins to develop. The embryo that forms is a miniature pine sporophyte, which lies dormant inside the seed until the time of germination.

- The cone scales finally open, and winged seeds are shed, generally during the third growing season for pines. And the process starts all over again.

5. The theory of life-cycle evolution states that an alternation of two equal generations in the same cycle is a "primitive" trait.

➤ **Taking It Further: Seed Plants**

1. *Answers will vary. The student is asked to list the evidence both for and against evolution from the fossil record. See the Supplement (pages 276–288).*

◆ **Think-Session Guide for Unit 8**

Plant Life

Subject: Light and plant growth

Purpose: To stimulate interest in plants

a. **To the student:** What effect does light have on the height a plant will grow?

Teacher: Suggest that the more light, the taller and healthier plants will be. List all three possibilities: increased growth, inhibited growth, no effect.

---

b. **To the student:** Considering the hypothesis—that light promotes growth—construct an "If ..., then ..." statement which will lead to an experiment.

Teacher: In an "If ..., then ..." statement, the *if* is followed by the hypothesis; the *then* is followed by the predicted results. Here it becomes: *If* light promotes growth, *then* the more light, the greater the height of the growing plants. You may want to practice using the other two possible hypotheses.

---

c. **To the student:** Using the "If ..., then ..." statement, plan a controlled experiment.

Teacher: If it does not come out, suggest the idea of sampling: that is, perhaps more than one kind of seed should be used—say corn and bean. This expands on the idea of the "experimental-control" pair.

---

d. **To the student:** With the "If ..., then ..." hypothesis in mind, can you predict the possible results and their interpretation?

Teacher: Suggest this form: "If plants grown in the light are (taller than) (shorter than) (the same height as) plants grown in the dark, we conclude that...."

It would be worthwhile to point out that our predictions are based on previous experience. Our actual results may be more complicated than our predictions. In this experiment, we assumed that the seed plants selected would respond in the *same* way.

This "Think-Session" can be used as a base for an interesting experiment using plants.

UNIT 9
Theories of Biological Change

CHAPTER 21
Weaknesses of Geologic Evidence
Text pages 291-303

Inception of Evolutionary Theory
Lack of Fossil Evidence to Support Evolution
Methods of Fossil Dating Inconclusive
Alternate Interpretation of the Fossil Record

◆ Suggestions for Motivation or Enrichment:

1. Report on any personal contact with fossils or with specimens that may be carefully inspected. Inquire of anyone who has visited the petrified forest of Arizona, the La Brea tar pits of Los Angeles, chalk hill fossil beds, archaeological diggings, or museum displays of ancient remains, asking about the uniqueness of the findings.

2. Display in parallel position a copy of the Babylonian "Epic of Creation—Enuma Elish" (from an encyclopedia or Internet source, such as <http://www.piney.com/Enuma.html>) and the biblical account of Creation from Genesis 1. Contrast the credibility of the two accounts, which both date from approximately the same time of recording.

3. Collect as many different pictures as possible of the Grand Canyon. Look up the meaning of the word *hydrodynamics* in a dictionary and describe its apparent connection to the formation of the famous canyon.

4. Look in a good dictionary for the prefixes *paleo-*, *meso-*, *ceno*, and *zo-*. Write out the basic meanings and several occurrences that seem to relate to biology. Which of the prefixes is also used as a word root?

◆ Suggestions for Supplementary Reading:*

1. Lamont, Ann. "Great Creation Scientists: Nicolas Steno: Founder of Modern Geology and Young-Earth Creationist." *Creation*, 23: 4, 47–49. To read this article online, visit <http://www.answersin-genesis.org/creation/v23/i4/steno.asp>.

2. Grigg, Russell. "Darwinism: It Was All in the Family." *Creation*, 26: 1, 16–18. To read this article online, visit <http://www.answersingenesis.org/creation/v26/i1/darwinism.asp>

---
* The listing of these suggestions does not necessarily imply endorsement of content.

3. Plantinga, Alvin. "When Faith and Reason Clash: Evolution and the Bible." *Christian Scholar's Review*, XXI: 18–33. To read this article online, visit <http://www.asa3.org/ASA/dialogues/Faith-reason/CRS9-91Plantinga1.html>. Not written from a young-earth perspective; clearly shows the fundamental religious nature of evolutionary theory and illustrates many of its weaknesses

4. Morris, John D. *The Geology Book.* Wonders of Creation Series. Master Books. <http://www.masterbooks.net>. Author explores the earth's crust, pointing out the evidences for creation (see ad for details of book)

5. Wofford, Wayne. "The Intelligent Design Movement." *Scientific Voice, Center for Scientific Studies.* To read this article online, visit the Union University Web site at <http://www.uu.edu/centers/science/voice/article.cfm?ID=22>. Introduction to the intelligent design movement

6. *Intelligent Design and Evolution Awareness (IDEA) Club.* To access the *IDEA Center* Web site, visit <http://www.acs.ucsd.edu/~idea/index.shtml>. This site (1) promotes, as a scientific theory, the idea that life was designed by an intelligence and (2) facilitates discussion, debate, and dialogue over these issues in a warm, friendly, and open atmosphere where individuals feel free to speak their personal views.

◆ Answers to Questions

Inception of Evolutionary Theory (Text page 292)

➢ **Questions: Inception of Evolutionary Theory**

1. Lamarck's and Darwin's theories of evolution:

	Lamarck	Charles Darwin
Conception of species:	Population of individuals all of the same kind (identical characteristics in all members). Individuals capable of transformation.	Population of interbreeding individuals with similar characteristics, though variation is common among all of them at all times. Individuals fixed and unchanging. Population capable of transformation.
Mechanism of new species production:	Internal drive toward greater complexity modified by the inheritance of acquired characteristics. Change directed to meet organism needs.	Natural selection. Variation exists regardless of organism's needs—not directed toward any purpose.

Example of this type of explanation (how the model accounts for some phenomenon):	The giraffe's neck: "At some point in the past, giraffes must have found themselves in an environment where they had difficulty reaching food present on the tops of trees. In order to eat, they must have had to stretch their necks and in doing so physically elongated them some. This longer neck was passed on to the offspring in the next generation, who in turn stretched their necks even further, thus resulting in the giraffe species having very long necks."	The giraffe's neck: "In each generation of giraffes the lengths of the necks would vary slightly. Those with longer necks would be able to reach higher for leaves. In the struggle for existence, the giraffes with the shorter necks would be less vigorous and thus produce fewer offspring because they could not get as much food as the animals with longer necks. The animals with longer necks would pass their traits on to their offspring, which would have longer necks than the previous generation. Long necks would have been such an advantage that the short-necked giraffes would eventually be eliminated. Natural selection would eventually result in greatly increased neck lengths."[a]
Phenomena the model can account for:	• Adaptation • Fossil record	• Adaptation • Fossil record • Homologous structures • Biogeographical diversity patterns

a. This excerpt is taken from page 292 of the text.

Teacher: This chart is largely based on the *Comparing Theories: Lamarck and Darwin* Web page found at the Science Netlinks Web site at <http://www.sciencenetlinks.com/lessons.cfm?BenchmarkID=1&DocID=387>. To download a PDF of their chart, click on the link **Lamarck and Darwin: Summary of Theories student sheet**, under the heading PLANNING AHEAD, and the subheading MATERIALS.

2. The idea of *mutations as the agent for change* was added to the theory of evolution through the study of genetics. Though Darwin was ignorant of any genetics, de Vries worked when genetics was becoming established as a science. He proposed that mutations provided the variations on which selective pressure could act to produce new species. It is now known that genes do not change except by mutation, which rarely occurs.

3. The burden of proof required to substantiate evolution is as follows:
   • It must be shown that one animal species is changing into another animal species; or that one plant species is changing into another plant species.

   • It must be shown that such changes have taken place in the past by tracing the so-called "tree of life" in the rocks.

In other words, it should be demonstrable in the laboratory and from the fossil record that life succession through genetically related organisms has passed through changes from the small to the large, from the less complex to the complex.

➤ **Taking It Further: Inception of Evolutionary Theory**

1. *Answers will vary. After doing some research, the student is asked to discuss the religious ramifications of the theory of evolution.*

Lack of Fossil Evidence to Support Evolution (Text page 297)

➤ **Questions: Lack of Fossil Evidence to Support Evolution**

1. The different types of fossils are as follows:
   • *The original substance of fossils might be preserved* (insects preserved in amber, plants and animals frozen, others preserved by drying or in wax or asphalt)
   • *Fossils might be replaced by another material* (replacements are with silica, calcite, or pyrite; usually, only hard parts of the organism are replaced)
   • *Other fossils may be only imprints* (tracks of crawling animals, wormholes, footprints, and imprints of other parts of the plant or animal body)

2. An *index fossil* is a fossil usually with a narrow time range and wide spatial distribution that is used in the identification of related geologic formations. Geologists use fossils to determine the relative ages of the rock layers; they assume that deep layers are older, as far as deposition is concerned, than the layers of overlying rock, unless the layers have been disturbed.

3. Reconstructing animals from fossilized remains is problematic for scientists because soft parts must be deduced. There is no way of knowing, for instance, the color of the animal, what kind of hair it had, or whether the ears were long or short. Thus, paleontologists have been known to make mistakes in trying to make restorations (e.g., drawings of the mammoth were different after the entire animal was found frozen in arctic muck).

Teacher: Several examples from the text are listed below, but the student is only required to give **five**.

• No attempt to restore some extinct fossilized organisms has been made because there is no living form with which they can be compared.
• Restorations are made of dinosaurs even though the paleontologists know much of the restoration is only what they have imagined.
• The family Fusulinidae is entirely extinct. They are classified into genera and species according to their structure, but no attempt is made to describe the animal that lived in the shell—no one knows!

- Another problem of interpreting fossils is that of altered structure due to disease or injury of the organism while it was alive.
- Not only do diseased individuals sometimes provide problems for the paleontologists, but also their size can be misleading.
- A consideration of living species provides another problem for the interpretation of fossils by the paleontologist.[*]
- Some animals are so different at different ages (e.g., tadpoles and adult frogs) that it is possible to classify some fossils as different species when actually they are juvenile and adult forms of the same species.
- The sexes of some kinds are very different. It is probable that, because of these differences, many fossils are erroneously classified.
- Hybrids may provide a basis for erroneous classification, also (e.g., the mule, a sterile hybrid between a female horse and a male donkey).
- Extinct animals might have produced sterile hybrids that are classified by paleontologists as species.
- Evidence there really is for the horse "series" is difficult to ascertain, because information seems unobtainable as to how many specimens of a supposed genus have been found and how complete they are.
- The fossils of these horses are found widely scattered in Europe and North America. There is no place where they occur in rock layers, one above another.
- It may be that the small size of some of the horselike animals was caused by poor feed. It may be that side toes were lost through mutation. Some of the fossils probably represent genera not related to the horse.

➤ **Taking It Further: Lack of Fossil Evidence to Support Evolution**

1. *Answers will vary. The student is asked to explain why and how science (particularly studies of extinct animals) is limited; he is also asked to give examples.*

Methods of Fossil Dating Inconclusive (Text page 300)

➤ **Questions: Methods of Fossil Dating Inconclusive**

1. *Uniformitarianism* is the principle that fossils and geologic strata were laid down by existing physical processes, but this is impossible since present-day processes simply could not have formed some features; thus geologists have to accept a certain amount of catastrophism. Most features can be better explained as being formed rapidly by great forces than by mild forces acting over vast lengths of time (e.g., the Grand Canyon[**]).

---

[*] For example, the skeletons of lions and tigers are so nearly identical that if they were found as fossils they would be classified as the same species. Yet the living animals are classified as different species.

[**] A better explanation is that the Grand Canyon was formed rapidly as water cut through not yet consolidated material that had been deposited by the flood of Noah's time, which conforms to the principles of hydrodynamics. These principles state that water cannot meander at the same time that it is cutting a deeper channel. Also the eruption of Mt. St. Helens points to catastrophism as the cause of canyon formation.

2. Some of the problems with the geologic timeline are as follows (*only five are required*):

- The Grand Canyon is considered a perfect example of strata in correct order. But the only "periods" represented there are Cambrian, Devonian, Mississippian (Carboniferous), and Permian. The Ordovician and Silurian above the Cambrian, and the Pennsylvanian above the Mississippian are absent.
- Not only is there no known place where all the "periods" or "epochs" are represented, but often they are in reverse order (e.g., the "Lewis Overthrust" in Glacier National Park in Montana).
- There are tremendous "time" gaps in many places. Many instances can be found where a "young" bed rests directly on a "very old" bed without evidence of an eroded surface between.
- Earth's surface is termed a "deceptive conformity" because the physical evidence contradicts the fossil evidence. In these cases, many geologists trust their preconceived theory rather than the physical data.
- The repetition of a layer several times in a section of earth is also difficult to explain. These things are hard to explain in terms of uniformitarianism.
- There is no example in man's knowledge of a geological upheaval big enough to account for the formations over large areas being in reverse order.

Teacher: While the student is only required to give the above answers, any of the following possibilities, found elsewhere in the chapter, may be used instead.

- The Precambrian is divided into the Archeozoic and Proterozoic eras. The former is represented by no fossils; the latter by only a few, all of which are disputed.
- In some cases, it is questioned whether certain fossils belong to the Precambrian or the Cambrian.
- In other cases, there is doubt as to whether they are really fossils at all.
- The slightly different nomenclature used in Europe is confusing since no clear international agreement has been reached regarding labels for divisions.
- "Eocambrian problem" is the sudden appearance of complex forms in the Cambrian.
- The above organisms have no fossil ancestors in the Precambrian; it is rather hard to explain why no fossils are recorded from these so-called "eras."
- Although fossils from the Cambrian period were different from species living now, they were not simple (e.g., trilobites, as complex as arthropods of today).
- One of the most striking things about the Mesozoic era (dinosaurs) is the sudden appearance of flowering plants just like present-day species. Yet flowering plants have no ancestors in the fossil record.
- When *Zinjanthropus* was first found he was hailed as "early man," but the fact that *Homo habilis* was found in deeper strata posed a contradiction to evolutionists in their idea of human evolution.
- The fossils found are almost always very fragmentary, and much imagination enters into each reconstruction of what these men, if they actually were men, looked like.

3. *Assumptions* involved in *absolute dating*:

- None of the daughter element was present in the rock when it was formed.
- The rate of decay of the element has remained constant since the time the rock was formed.
- All the daughter element in the rock was derived from the parent element that was previously in that rock.

Problems involved in this type of dating:

- The reverse is possible: some of the parent or daughter element may have escaped from the rock.
- It has been shown that cosmic radiation can alter decay rates.
- If the amount of cosmic radiation has varied in the past, these dating methods would be invalid.

➤ **Taking It Further: Methods of Fossil Dating Inconclusive**

1. *Answers may vary. The student is asked to explain why a system as arbitrary as the geologic timeline would become so widely used.*

Alternate Interpretation of the Fossil Record (Text page 303)

➤ **Questions: Alternate Interpretation of the Fossil Record**

1. Evolutionists tightly adhere to the idea of an old earth because they need eons of time as a framework for the doctrine of evolution; thus, they reject dating methods by which the age of the earth is interpreted as quite young

2. Some counter examples mentioned in this chapter that suggest errors in the evolutionary theory are as follows:

Teacher: Many examples from the text are listed below, but the student is only required to list *five*.

- "Gemmules," natural selection, and mutations are inadequate mechanisms to cause major changes that can be inherited and cause an evolution to a more complex and desirable condition.
- As for giraffes, the slight differences in animals' necks is now known to be caused by differences in food, or by possession of a different number of the dominant genes that control the neck length. Mutations for greater neck length have never been observed.
- The most striking case of faulty interpretation due to disease is Neanderthal man. Later, it was discovered that this particular individual had osteoarthritis that bent his back into the curved shape pictured in the cavemen restorations.
- Extinct animals may have produced sterile hybrids that are classified by paleontologists as species (e.g., the possible case of the horse series and the sea urchin series in England).
- Today there are listed at least twenty-six genera of horses, of which no biological changes exemplifying the doctrine of evolution (change of one form, or kind, into another) have been found. Consequently, so-called horse evolution is actually no more than

possible variations within describable limits of the horse form, or kind.

- The fossils of these horses are found widely scattered in Europe and North America. There is no place where they occur in rock layers, one above another.
- Although fossils from the Cambrian period were different from species living now, they were not simple (e.g., trilobites, as complex as arthropods of today).
- One of the most striking things about the Mesozoic era (dinosaurs) is the sudden appearance of flowering plants just like present-day species. Yet flowering plants have no ancestors in the fossil record.
- When *Zinjanthropus* was first found he was hailed as "early man," but the fact that *Homo habilis* was found in deeper strata posed a contradiction to evolutionists in their idea of human evolution.
- Not only is there no known place where all the "periods" or "epochs" are represented, but often they are in reverse order (e.g., the "Lewis Overthrust" in Glacier National Park in Montana).
- Many instances can be found in Canada and the United States where a "young" bed rests directly on a "very old" bed without evidence of an eroded surface between.
- The shellfish *Lingula* is found in fossils formed in Ordovician rock layers to the present, but there is virtually no change over an estimated period of some 500 million years.
- Likewise, the sequoia trees of California have not evolved over the years.
- The opossum is an organism that we might expect to be extinct, because it has a small brain and is not specialized in its teeth, feet, or legs. Yet, it has extended its range from Middle America into New England and has become abundant in California.
- A most spectacular example of a fossil in the "wrong" formation is the sandal-like print with several trilobites in it (found June 1, 1968, by William J. Meister, near Delta, Utah). From this, one can deduce that man and trilobites lived at the same time. Yet, according to present geological theory, trilobites were extinct hundreds of millions of years before man came on the scene.
- Another spectacular find was that of giant men's footprints in a Cretaceous riverbed near Glen Rose, Texas. Also found in the same bed were dinosaur and brontosaurus tracks. Clearly, this makes possible the deduction that man and the dinosaur lived at the same time. Men were not "supposed" to have existed in the Cretaceous period, or for sixty or seventy million years.

3. The Green River Basin (Eocene) of Colorado and Wyoming is considered part of an old lakebed. Masses of fish fossils are found in this formation, far more than can be accounted for by present day processes. On the basis of the excellent preservation and large numbers, a quick burial, or worldwide flood, is the most logical explanation for large beds of fossils.

➤ **Taking It Further: Alternate Interpretation of the Fossil Record**

1. *Answers will vary. The student is asked to explain what role the media plays in introducing bias regarding the theory of evolution.*

➤ **Questions: Chapter Review**

1. *Evolutionists*: George L. Buffon, Erasmus Darwin, Charles Darwin, Francis Galton, E. Geoffrey Saint-Hilaire, Jean-Baptiste de Lamarck, Hugo de Vries, G. A. Kerkut, Louis S. B. Leakey, Sir Charles Lyell.

   Creationists: Carolus Linnaeus, George Cuvier, Gregor Mendel. Prominent creationists not mentioned in this chapter were Louis Agassiz and Henri Fabre. Some of the evolutionists mentioned facts which favor creationist belief (e.g., Galton).

2. *Use inheritance, environmental inheritance*, and *inheritance of acquired characteristics* are the same belief. It is claimed that a change in a living thing caused by the environment or by use will appear in the next generation in the absence of the causative agent. Experiments have failed to give any positive results. Gene theory also does not support this belief (see chapter 7).

3. An *index fossil* is supposed to have come into existence within a certain period and to have become extinct within a certain known period. Its presence in a rock is supposed to indicate the time within which that rock was laid down. Evolutionary geologists rely on index fossils more than anything else for dating other fossils from similar rock layers.

4. Some of the sources of error in identifying fossils are erroneous dating methods, poor preservation, breakage, and the wear of long transportation.

5. *The answer is not explicitly found in the text but may be deduced from the context.*

   Darwin's explanation of the inheritance of acquired characteristics is as follows: "gemmules" in the bloodstream are affected by the new character and inherited by the next generation. Galton injected blood into rabbits and disproved the "gemmule theory" (actually a *hypothesis*). Yet, Darwin is so highly esteemed that he is mentioned and Galton is passed by.

6. Shells found on a beach that are washed back and forth by the waves would **not** make good fossils because they are rubbed against each other by the waves until each becomes a smooth piece of mother-of-pearl and their species characteristics are lost. In contrast to beach shells, many fossils have all the sharp and fragile points and edges preserved. This condition indicates that they were not washed about but were covered quickly.

7. George Cuvier was one of the first scientists to point out this difference between fossils and present beach shells. This is evidence for cataclysm (catastrophism) rather than slow changes as at present (uniformitarianism).

Teacher: The following question is prefaced by these comments: *To explain the lack of intermediate kinds of fossils some geologists suggest that the animals changed quickly from one kind to another. Thus, there were few intermediate fossils, and they have not been found.*

---

8. According to Charles Lyell, all changes have been slow and regular—no faster than are observed today. Evolutionists are supposed to agree with Lyell, yet they can and do suggest other occurrences to extricate their theory of evolution.

9. If there was excellent proof of evolution, then plants and animals below the Cambrian series of rocks should be abundant and only a little simpler than the abundant ones in the Cambrian. You can see that this expectation is defied by the fact that nothing is found in this sub-Cambrian layer.

CHAPTER 22
Evidences From Similarities
(Text pages 305-312)

Structural Similarities
Developmental Similarities
Biochemical Similarities
Misapplication of Similarities

◆ **Suggestions for Motivation or Enrichment:**

1. Examine some small vertebrates, such as fish, bird, dog, and cat; analyze the forelimb adaptations as contrasted with the human forelimb. Select one type of animal and prepare a specific report on the structure and functioning of its forelimbs.

2. From personal knowledge, discuss if the surgical removal of apparently unnecessary parts of the human body is helpful or detrimental. Hypothesize on the original use of those parts, including tonsils, appendices, gall bladders, and spleens.

3. From personal knowledge or from pictures prepare a list of outstanding examples of protectively colored animals.

4. Compare the Bible references of Genesis 7:2 and Leviticus 11. Make two lists of animals that Noah had to consider for the ark. Remember that Noah must have had personal knowledge of what was *clean* and *unclean* in his time. Also, discuss the similar distinctions in the modern Jewish observance of Kashruth (dietary laws). Talk to someone who has traveled or lived under such observances.

◆ **Suggestions for Supplementary Reading:**[*]

1. Wieland, Carl. "The Moth Files: An Update on the Peppered Moth Fiasco." *Creation*, 25: 1, 14–15. To read this article online, visit the *AiG* Web site at <http://www.answersingenesis.org/creation/v25/i1/moth.asp>.

2. Frair, Wayne. "Embryology and Evolution." *Creation Research Society Quarterly*, 36: 2, 62–67. To read this article online, visit <http://www.creation-research.org/crsq/articles/36/36_2/embryology.html>. Criticizes the idea of recapitulation as a support for evolution

◆ **Answers to Questions**

Structural Similarities (Text page 306)

➢ **Questions: Structural Similarities**

1. The general principle that scientists follow when looking for evidence of evolution is actually an *assumption* that the degree of similarity of organisms indicates the degree of supposed relationship of said organisms.

2. *Homologous structures* are important to evolutionary theory because evolutionists believe these structures show common ancestry of animals.

➢ **Taking It Further: Structural Similarities**

1. *Answers may vary. The student is asked to explain what a bias is and how it affects the work of a scientist.*

Developmental Similarities (Text page 308)

➢ **Questions: Developmental Similarities**

1. "Ontogeny recapitulates phylogeny" is the idea that the embryo of an organism goes through stages of embryonic development wherein stages of its evolution could be detected.

2. A *vestigial organ* is a structure for which no use has been found and, therefore, is supposedly obsolete.

3. Some problems with both the supports of evolution mentioned above are as follows:

Teacher: Several problems from the text are listed below, but the student is only required to give *five*.

- **Human heart**—supposedly, the heart passed through a worm, fish, frog, and reptile stage before reaching its final form. The heart in the human embryo has *one chamber* (as in the worm), *two chambers* (as in the fish), *three chambers* (as in the frog), and *four chambers* with a connection of the two sides (as in the reptile). But the heart in man starts out with two chambers, which fuse into one for a time. This sequence actually reverses the stages of supposed evolution.
- **Structures resembling fish gills**—supposedly, these "gills" show a fish ancestry for human beings. Actu-

ally, they are alternating ridges and furrows on the right and left sides of the neck. They *never* develop into gills. They remain covered by a thin membrane and *never* have a respiratory function.

- **Primitive kidneys**—supposedly, these "worm and frog stage kidneys" show a worm/frog ancestry for man but do not function in excretion at any stage.
- **Appendix**—supposedly vestigial, evidence now shows that man's appendix serves as an aid in the body's defense against disease. Since fossils of ancient man consist only of bones, there is no way to determine what early man's appendix was like.
- **Nictitating membrane**—supposedly vestigial, its main use is to collect foreign material that gets in the human eye and deposit it in a sticky mass in the corner of the eye where it can do no damage.
- **Coccyx**—this supposedly vestigial "tail" temporarily protrudes during embryonic development; but it does not shrink, but becomes surrounded by the developing hips and serves as a place of attachment for certain muscles—a very important function for good posture.
- **Endocrine glands**—supposedly vestigial, they have great importance as hormone producers.
- **Thymus gland**—supposedly vestigial, it has recently been found to be involved in protecting the body from disease.

➢ **Taking It Further: Developmental Similarities**

1. *Answers may vary. The student is asked to discuss several misconceptions in science that were later refuted by further investigation.*

Biochemical Similarities (Text page 310)

➢ **Questions: Biochemical Similarities**

1. Some biochemical similarities between mankind and other organisms are as follows:

- The nucleic acid DNA (deoxyribose nucleic acid) is found in the nuclei of all living things.
- All the genetic "information" necessary for the production of each particular "kind" of man, animal, or plant is found "encoded" in the DNA.
- All such variations in traits are implicitly present in the genetic material at the beginning of each individual organism. There is no way for truly new traits to appear except by mutations, and these are nearly always harmful.
- ATP (adenosine triphosphate) is universally involved in energy transfers of organisms.
- Hormones and enzymes in man and various mammals are so much alike that animal hormones are used to treat human diseases.

2. A *new trait* is an inherited characteristic that a particular organism did not have. There is no way for truly **new** traits to appear except by mutations, and these are nearly always harmful.

3. *Protective coloration* is coloration that enables an organism to escape its predators. It has been used to support the theory of evolution as follows: since the earliest birds or other organisms were of various colors, supposedly predators caught the ones

---

[*] The listing of these suggestions does not necessarily imply endorsement of content.

that could best be seen and overlooked the protectively colored ones, which are said to have had a "selective advantage." However, such an inference from presumptive evidence is not verifiable.

➤ **Taking It Further: Biochemical Similarities**

1. *Answers may vary. The student is asked to list the ways in which biochemical similarities support evolution and ways in which they support special creation.*

Misapplication of Similarities (Text page 312)

➤ **Questions: Misapplication of Similarities**

Teacher: Note the following answer is inferred, not directly stated, in this section; however, chapter 9 (section "9-5 What is a Species?") answers this question.
---

1. The difference between the terms *kind* and *species* is that a *kind* refers to broad group of creatures, such as cats and dogs (not necessarily *species*), that were created with similarities and differences; a ***species*** (Latin for "specific") refers to a group of similar, related organisms capable of breeding together and distinct from other groups.*

Teacher: At first, Linnaeus believed that the various *species* of organisms were created separately at the beginning; but later in life, he proposed a modified view, believing that the created *kinds* were capable of great variation. This is the creationist view today.
---

2. *Circular reasoning* is a fallacy in reasoning in which the premise is used to prove the conclusion, and the conclusion used to prove the premise. For example, organisms are arranged on a "phylogenic tree" as the evolutionist thinks they evolved, and then the fact that selected organisms can be so arranged is considered as evidence that they did evolve. Logically, this is known as a *vicious circle*.

3. *Answers may vary in regard to which means of classification is better (DNA or physical structures). However, the book does state the following:* Classification by either DNA or physical structures is obviously better than the evolutionist idea of classifying according to presumed evolutionary history. Classification would then be based on empirical evidence that can be verified rather than on theories of history that cannot be investigated in the science laboratory or in the natural environment of organisms.

➤ **Taking It Further: Misapplication of Similarities**

1. *Answers may vary. The student is asked if the absence of intermediate fossils is evidence that evolution did not occur and to explain why or why not.*

➤ **Questions: Chapter Review**

1. A vestigial organ would not be functioning in the present generation, therefore it would be useless

---
* Evolutionists believe the characteristics resulted from evolutionary changes.
---

and a loss. It is evident that vestigial organs do not give evidence for evolution, which postulates upward changes of the most magnificent scope.

2. The following quotation is a sample of the suppositions:

> The monarch is a very distasteful mouthful to birds and other insect eaters because of the acrid taste, and so flies freely about fields and gardens, seeming to sense this protection.
>
> The viceroy is unusually tasteful, but because of its striking resemblance to the monarch it, too, is let severely alone.**

If monarchs had been fed to birds and the birds had become sick, the evidence might be scientific.

3. Squids and octopods are classified as mollusks, yet their eyes are very similar to those of vertebrates. The presence of eyes could be taken as a basis of classification instead of the presence of vertebrae, and then squids and horses would be in the same phylum. If the color of the blood were taken as the criterion, earthworms would be added to this large group. If we say that classification proves kinship, we run into difficulties. Rather, it is a man-made device of convenience.

4. It seems that the light strain of the peppered moth is the original, and the dark strain is a mutation. The reproductive rate of the original strain is the higher; but the dark strain was developed during the "Industrial Reformation"*** in Great Britain when soot from factory chimneys turned the light-colored bark of the trees nearly black. This suggests that industrial pollutants could have induced the mutations in the peppered moth.

Despite the conditions favoring the dark strain of the peppered moth, the light strain still survived. When the tree bark was lightened due to less coal pollution, the population of the light strain of the peppered moth increased.

5. Evolutionists say that man is most closely related to the animals which have most nearly the same structure. They say also that kinds of animals which have the same parasite are most closely related. The two statements do not agree.

6. There is lack of observation of one kind of organism giving rise to another of more complex structure. This is what a "trunk" organism would have to do. Estimates have been made as to which group of plant or animal is ancestral to another, but there is such lack of agreement in these estimates that now an ancestral tree is often drawn without a trunk.

** Lilian D. Frazinni, Williams Pub. Co., 1934, p. 20.

*** Garry J. Moes, *Streams of Civilization, Volume Two* (Arlington Heights, IL: Christian Liberty Press, 1995), pages 153–160.

7. The heart of a vertebrate develops from a muscular tube to a four-chambered organ. Certain adult animals can be listed in a sequence from those with simple hearts to those with complex hearts, but the development of an embryo does not proceed in this order. The heart starts with a tube which is double, later changing to single tube, then to the heart with chambers.

CHAPTER 23

Early Man

Text pages 315-321

Classification and Uniqueness of Man
Search and Discovery of "Early" Man

◆ Suggestions for Motivation or Enrichment:

1. Discuss the concept "A dog is man's best friend" as a commentary on the evolutionist's elevation of apes and monkeys to man's closest animal connection.

2. Research some of the fascinating modern findings in animal communications, such as with porpoises, pigeons, and horses. Discuss the apparent limitations of such communication.

3. Inquire of some anthropology resource person or textbook about the concept of man's algebraic mentality. Speculate about what an ape might be like who showed such mentality.

4. Illustrate from a few different languages the symbolism required for human language and the abstractions which can be revealed in words. Prepare a brief presentation on the development of Chinese or Native American picture writing.

◆ Suggestions for Supplementary Reading:*

1. Mehlert, W. (Bill). "*Homo erectus* 'to' modern man: evolution or human variability?" *Creation*, 8: 1, 105–116. To read this article online, visit <http://www.answersingenesis.org/tj/v8/i1/erectus.asp>.

2. Gish, Duane T. "Man ... Apes ... Australopithecines ... Each Uniquely Different." *Impact*, November 1975. To read this article online, visit <http://www.icr.org/pubs/imp/imp-029.htm>.

◆ Answers to Questions

Classification and Uniqueness of Man
(Text page 316)

➤ Questions: Classification and Uniqueness of Man

1. The similarities and differences between men and apes are as follows:

- **Similarities**—men and apes both have hair, females have mammary glands, physical features are similar (shape of the face, nails on fingers and toes); larynx is similar (but ape cannot be taught to speak);
- **Differences**—men are so much more intelligent than apes; ape cannot be taught to speak; physical differences (foot of an ape fitted for clasping tree branches, rather than walking); apes usually walk on all four limbs, the arms being as long as the legs; hair is evenly distributed over the body of apes; different bone structures—skulls, brows, and teeth; behavioral differences; man has language, reasoning, and organizational skills that apes do not have

2. *Answers may vary. The student is only required to give* **two**. Some creatures are similar in structure but otherwise very different organisms:

- Vertebrates are placed in the same group based on their *backbone* (dorsal series of bones); but, if *hemoglobin in the blood* were the criterion, the earthworm would be included in this group.
- The crayfish and grasshopper are placed in the same phylum because both have an *exoskeleton*; but, if they were classified on the *method of carrying oxygen*, they would be in separate phyla—the crayfish carries oxygen in the blood, whereas the grasshopper uses a system of tubes called tracheae.
- Men and apes are similar in structure, but there are many differences: a huge gap in intelligence; man's exclusive use of language; physical differences—their skulls, feet, limbs, etc.; and behavioral differences—apes cannot make tools, organize to defend themselves, or carry wounded in retreat.
- The squid has eyes very much like the vertebrates, but no one claims that they are closely related.

➤ Taking It Further: Classification and Uniqueness of Man

1. *Answers may vary. The student is asked to find some current articles on the evolutionary stages connecting man with lower animals and summarize the assertions made.*

Search and Discovery of "Early" Man
(Text page 321)

➤ Questions: Search and Discovery of "Early" Man

1. Finding the remains of Cro-Magnon man and Neanderthal man together infers that:

- They were probably living together or, at least, in the same region, based on remains found in Israel.
- They stood tall and erect, their bones were strong, and their brains were as large as, if not larger than, that of modern man.**
- They must have been very intelligent and skilled artisans because of the tools, artifacts, paintings, and carvings found near their various remains.

---

* The listing of these suggestions does not necessarily imply endorsement of content.

** Neanderthal man's cranial capacity averaged 1,450 cc, and that of men today is usually no larger; Cro-Magnon man's cranial capacity was about 1,660 cc, which is considerably larger than that of modern man.

- They apparently defended themselves with weapons, used fire, and buried their dead with the things that they thought would be useful in the afterlife.
- They may have come after modern man, since remains of two human skulls were found in a deeper layer than the tools made by Neanderthal man.

2. *Answers may vary. Some of the information has been added from outside sources. The student is required to discuss only **five** of the following answers.*

For early men mentioned in the text, a list of their assigned genera or species, where they were found, and their characteristics are as follows:

- Neanderthal man—*Homo sapiens neanderthalensis*; they were found in the Neander valley near Dusseldorf, Germany (also Europe and western Asia); very similar anatomically to modern man (*H. sapiens*)*
- Cro-Magnon man—"early" man (*H. sapiens*); found in Les Eyzies in France; tall and well-formed, created tools and art, with skull slightly larger than modern man's; Creationists believe: a true human—post-flood descendant of Noah

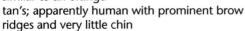

- Wadjak Man—classified by Dubois as *Pithecanthropus erectus* ("ape-man who stands erect"); found near Wadjak, Java (Indonesia); a flat, very thick skullcap, and its teeth are very similar to an orangutan's; apparently human with prominent brow ridges and very little chin
- *Paranthropus robustus*—found near Swartkrans (South Africa); similar to modern man but smaller
- Africa man—*Zinjanthropus*; found in Kenya, East Africa; teeth like modern man, but has crest projecting from the upper part of the skull, which is not as prominent as that of the gorilla
- Able Man—*Homo habilis*; found by the Leakeys at Olduvai Gorge in Tanzania; the skull, hands, and feet are very similar to those of modern man, but he is only four feet tall
- Swanscombe Man—archaic *Homo sapiens*; found in Kent, England; bones are very thick, with a mixture of primitive and modern features, and an estimated brain size of 1325 cc.
- Peking Man—*Sinanthropus* (mentioned only in passing on page 315 of the text); found near Beijing (formerly Peking), in China; apelike features according to Gish (1985)

Teacher: *Australopithecus garhi* is a new species of early hominid from Ethiopia; your student may want to do further research on this recent find (1996). Visit <http://www.mnh.si.edu/anthro/humanorigins/whatshot/1999/wh1999-1.htm> for more information.

➤ **Taking It Further: Search and Discovery of "Early" Man**

1. *Answers may vary. The student is asked to list some ways to prevent bias from entering into scientific research.*

➤ **Questions: Chapter Review**

1. An animal is taken as a type of a large group and studied carefully in the laboratory. Both time and interest would run out if an individual made a careful study of each species.

2. Animals are classified according to their structure (morphology). In his structure, man is more like the apes than like any other creature. Similarity of structure, however, does not prove descent from a common ancestor.

3. *Answers may vary. The student will need some outside source, such as the Internet or library, to answer this problem.***

For forty years the fragment of jaw and a part of a skull were considered one of the archaeological finds of the century: proof that man evolved from the apes. They were the bones of *Eoanthropus dawsoni* found near Piltdown Common in Sussex. The cranium was that of an undoubted man, while the lower jaw was from a recent ape. It took scientists forty-one years to discover the fraud. Piltdown man had been widely publicized as a "missing link."

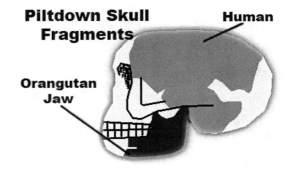

4. Physical differences between man and ape are seen in the teeth, shape of jaws, size of brain, ridges above the eyes, pelvic girdle, shape of feet, and relative length of arms and legs.

---

* They are so similar, in fact, that in 1964 it was proposed that Neanderthals are not even a separate species from modern humans, but that the two forms represent two subspecies: *Homo sapiens neanderthalensis* and *Homo sapiens sapiens*.

** Richard Harter's Web page, "Piltdown Man: The Bogus Bones Caper," is an excellent online resource. (Visit <http://www.talkorigins.org/faqs/piltdown.html>. This pro-evolutionary Web site tries hard to discredit scientific creationism.)

These groups differ even more in spiritual and mental traits. Man uses fire, makes tools, buries his dead, has some form of religion, uses words as symbols of ideas, and profits not only from his own experience but also from that of others.

5. *Answers may vary. The student will need some outside source, such as the Internet or library, to answer this problem.* In 1939, Alberto Carlo Blanc and Sergio Sergi reported in *Science* the finding of a Neanderthal skull in a cave at Monte Circeo in Italy. Well-preserved skeletons of Neanderthal man showed that the skull was carried above the neck vertebrae as in man, not in front of the neck as in apes.[*]

6. Animals give cries of warning, of cheer, and probably of praise. They do not use words as symbols of ideas nor put words together in sentences. The action of one animal is often mimicked by others.

7. Eugene Dubois did not set out to learn about early man but to demonstrate that the guesses of Ernst Haeckel were correct. He did not have the true spirit of the scientific method, which is to follow truth wherever it may lead. Dubois had a common human weakness, attempting to prove that which he wanted to believe. All of us should be on our guard against such subjectivity.

8. Some persons believe that evolution has been amply demonstrated to be true. When a skeleton of low type is found, they jump to the conclusion that it is ancestral to modern man. Such persons forget that they are using their assumption of evolution as proof of evolution (circular reasoning).

CHAPTER 24
Problems for Evolutionists
(Text pages 323-328)

Mechanism and Evolution

Origin and Evolution

Transition and Evolution

◆ **Suggestions for Motivation or Enrichment:**

1. Evaluate albinism in human mutation, as related to eyesight, sunburn, hair styling, and pagan superstitions. Discuss any other known mutations in animals or plants.

2. Postulate a group of apes facing a critical need for crossing a sea. Describe their limitations in the areas of group planning, imaginative construction, and self-denying work.

3. Research for information on "Thalidomide babies" and contrast the explanations with the concept of mutation.

4. Recall evidences of long-range working of conscience, as in news reports of anonymously returned money, years after a theft. Try to account for the emergence of such conscience by stages of evolutionary development.

◆ **Suggestions for Supplementary Reading:**[**]

1. Richards, Jay. 2002. "Not by Chance: Darwinism, Intelligent Design and the Materialistic Worldview." *American Family Association Journal*, August 2004, 19–20. To read this article online, visit the *FAFJ* Web site at <http://www.afajournal.org/2004/august/804not_chance.asp>. This article is an introduction to the intelligent design movement and its challenge to Darwinism.

2 Morris, Henry M. 2002. "The Scientific Case Against Evolution." Institute for Christian Research. To read this article online, visit the *ICR* Web site at <http://www.icr.org/bible/tracts/scientificcaseagainstevolution.html>.

◆ **Answers to Questions**

Mechanism and Evolution (Text page 325)

➤ **Questions: Mechanism and Evolution**

1. The current evidence about mutations being the mechanism of evolution implies that such a process seems insurmountable. The changes produced by mutations are in the wrong direction to support evolution because only one in a thousand examples may be a beneficial mutation. The net progress is downward, not upward. Moreover, mutations in all those organisms filling the various niches would have to be timed so that the entire ecosystem would evolve as a unit. Finally, *artificial mutations* produced by man-made changes in environment, such as some form of radiation, always jeopardize the viability of an organism.

2. The various problems with intermediate stages are as follows:

- If a new structure is started but not well enough formed to be functional, that structure is a hindrance rather than a help (e.g., the supposed evolution of the vertebrate eye—intermediate stages would be useless to the survival of an organism).
- Establishment of life processes is even harder to visualize by chance variation (e.g., the nucleus of a cell divides by such a precise process that each daughter nucleus has a copy of each chromosome).
- The arrangement of chromosomes, genes, and DNA codons is so complex and so well structured that any improvement from chance accidental changes is hard to imagine.

---

[*]   Visit <http://www.custance.org/Library/Volume2/Part_V/Chapter2.html> for information.

[**]   The listing of these suggestions does not necessarily imply endorsement of content.

➤ Taking It Further: Mechanism and Evolution

1. *Answers may vary. The student is asked to think of any intermediate stage between two species that would be beneficial to the organism.*

2. *Answers may vary. The student is asked to explain what the benefits of natural selection are.*

Origin and Evolution (Text page 326)

➤ Questions: Origin and Evolution

1. The currently held theory of evolutionary origin is that the origin of life occurred through *mechanistic processes*, following physical laws; but no such physical processes are known. Evolutionists are forced to abandon the law of biogenesis and advocate some theory of spontaneous generation of life, hoping that an answer will be found at the submicroscopic level of organization of life.

2. The problems with Oparin's theory are as follows:

   • Under the same conditions postulated by Oparin for an ancient atmosphere, there would have been no protection against excessive ultraviolet radiation. Oxygen, primarily in the form of ozone, in the present earth's atmosphere shields the earth from lethal dosages of ultraviolet light. If the Oparin process had been successful in generating life, ultraviolet radiation would have killed it.

   • Had Oparin postulated oxygen in the atmosphere, his theory would still have been in trouble because oxygen can be toxic to some cells. But the enzymes would not have "evolved" before the cells. Either way, *with* or *without oxygen*, this theory of the origin of life cannot survive the tests of scientific analysis.

➤ Taking It Further: Origin and Evolution

1. *Answers may vary. The student is asked to list the problems life would have in an atmosphere without oxygen, besides lacking a protective atmosphere. Given the chemicals present in Oparin's described early atmosphere, the student should state if there is any way to accommodate these problems.*

Transition and Evolution (Text page 328)

➤ Questions: Transition and Evolution

1. The fossil record seems not to support the theory of evolution for the following reasons:

   • Total absence of types considered to be most primitive and ancestral to animals without backbones
   • The sudden appearance of the major divisions of organisms
   • An amazing absence of any transitional forms

2. The theory of evolution contradicts the law of biogenesis in the following ways:

   • Controlled experiments show that different degrees of variation occur *within* possible basic forms, or kinds, of organisms but *never across* these basic divisions of plants or animals.
   • From these experiments come observations that are in total agreement with the law of biogenesis (life begets life) and in total disagreement with the theory of evolution.

➤ Taking It Further: Transition and Evolution

1. *Answers may vary. The student is asked to explain if it is more logical to believe that there is no transition fossil in existence or that it simply has not been found yet. He should state which idea requires more faith and support his answer.*

2. *Answers may vary. Studies done with primates to teach them sign language have raised questions about possible "human" characteristics being present in these mammals. The student is asked to find an article pertaining to this subject and summarize the findings.*

➤ Questions: Chapter Review

1. Wilhelm Johannsen of Denmark was able to sort out strains or pure lines of beans by selection. He was not able to make further changes, however. Selection within a pure line did not change future generations.

2. If all the genes of one individual of a group are like the genes of any other individual of that group, selection of any kind will not produce a change. This is the situation which Johannsen found.

3. Seedless oranges, stringless beans, hornless cattle, etc. are helpful to man but not to the organism which has the peculiarity. Only a few of these mutations have been claimed to be helpful except under changed conditions of the environment.

Teacher: Dr. T. Dobzhansky said, "If you mean changes which are helpful under the same conditions of environment, don't look for them."

---

4. Such examples of mutation as mentioned by this author have never been observed. The reverse kind, resulting in weakness or loss, are observed regularly. Another objection is that mutations occur sporadically rather than in series, as would be necessary to develop rabbits which could run faster or discover danger more efficiently.

5. Shifts in the average often have been mistaken for real change. Suppose that tall peas have been mixed with short peas. Seed is saved only from plants that grew more than two feet tall. The next year, the average height will be greater but no individual plant will have been changed.

6. Both natural selection and selection by man tend to maintain a standard by eliminating weak and sickly individuals. They make no biological change in any individual. The plants or animals which are "selected" to reproduce are not themselves improved by being selected. In other words, there has been no change in their genes.

7. If the doctrine of evolution were true, it would favor heartless thugs such as terrorists and weeds. An altruistic person would be less "fit" to survive.

On the other hand, where a majority of a group of people recognize God, they appreciate and favor the altruistic person.

CHAPTER 25
Limited Variation Versus Unlimited Change
(Text pages 331-337)

Genetic Variations
Natural Selection and Complexity
Conclusions

◆ **Suggestions for Motivation or Enrichment:**

1. List the characteristics by which you recognize people's race. Include hair, skin, eyes, nose, mouth, stature, and any abstract conceptions which seem to be important to you. Evaluate the latter in a short essay.

2. Identify a Pygmy living in the bush (equatorial Africa) and a well-educated professor living in an urban center (continental Europe) as of the same biological species. Try to devise one I.Q. test which would be fair for each man.

3. Visit a plant nursery, farm, or agricultural center and learn something of the techniques for safeguarding plant strains against interbreeding. In discussion with other students, share any personal experiences pertaining to plant mixtures.

4. Research reference sources on selective breeding of animals, such as race horses, beef or milk cattle, and domestic pets. By way of contrast, call or visit the local city pound and inquire about the breeding problems caused by unrestrained dogs and cats. In a short essay, present the pros and cons of stiffer laws, including the enforcement problems.

◆ **Suggestions for Supplementary Reading:**[*]

1. Bergin, Mark. "Unfashionable Genes." *World Magazine.* To read this article online, visit <http://www.worldmag.com/displayarticle.cfm?id=9734>. The author writes about the controversy that erupted when an intelligent design scientist had his work published in a scientific journal.

◆ **Answers to Questions**

Genetic Variations (Text page 334)

➤ **Questions: Genetic Variations**

1. *Answers may vary. The student is only required to give **five**; many more answers are possible. Some examples of variations within a species are as follows:*

- Pink Grootendorst—*Rosa rugosa*. Many varieties of roses exist today because of man's efforts in plant breeding.
- Maize or corn—*Zea mays*; sweet corn, popcorn, and field corn
- Red and white onions
- Yellow and white peaches
- Dogs—Pomeranian (small, compact, long-haired), Chihuahua (small, but with very short hair)
- Peppered moth—*Biston betularia*; dark and light species by natural selection

2. A *polyploid organism* is an organism where **three** or more **monoploid** sets of chromosomes are together in the cells of an organism. In most organisms, the gametes are characterized by a single, or monoploid, set of chromosomes. Ordinarily, when two gametes unite to form a zygote, then the new cell contains the **diploid** set of chromosomes, one set from each parent. Some plants have variations of three or more homologous sets of chromosomes, and this condition is called **polyploidy**. It must be remembered that most polyploid organisms have a reduced viability.

3. The finches in Darwin's study provide evidences, pro and con, for evolution as follows:

- **Supportive evidence:**
(1) Labels aside, if the birds were arranged according to body and beak size, a perfect gradation would be found between the species, from the species with the largest beak to the species with the smallest beak.

(2) Some individuals of *G. scandens* are identical with those having the longest beak in the *G. conirostris* variety that intergrades with *G. magnirostris*.

- **Contrary evidence:**
(1) Most specimens of *G. scandens*, which have a very long beak, are truly distinctive (they eat cactus fruit).

(2) These birds are an example of genetic principles. If a few animals are isolated, "latent" genes will be expressed and various characteristics will develop.

(3) Were it not for the historical value placed on these birds, it is doubtful if they would still be retained as true species.[**]

➤ **Taking It Further: Genetic Variations**

1. *Answers may vary. The student is asked to explain how continuing to classify the different finches in separate genera is an example of bias.*

* The listing of these suggestions does not necessarily imply endorsement of content.

** Note also that all types of human beings are assigned to one species and yet display even wider variation than Darwin's finches, classified as several species.

Natural Selection and Complexity (Text page 335)

➤ **Questions: Natural Selection and Complexity**

1. Different environments can make organisms, which are genetically identical, apparently different in the following ways:

   - *This example was mentioned in the previous section.* If a few animals are isolated, "latent" genes will be expressed and various characteristics will develop. This happens because the animals mate with others that also have the latent genes. If the finches had stayed on the mainland of South America, new characteristics of color and beak shape might not have been fully expressed.
   - If the genes are alike, as in identical twins or in a pure line of beans, any difference between the individuals would be due to environmental factors, such as quality and quantity of food, air and water quality, radiation, or disease. Such differences are not passed on to the next generation; therefore, they do not have genetic significance.

2. It is interesting that the opossum is expanding its range because it is an animal that is not specialized in teeth, legs, or brain, yet it is successful. Moreover, it is considered to be stupid, yet the opossum—more than the skunk, Arctic hare, porcupine, red fox, coyote, or gray wolf—has increased its range from the Middle Atlantic States into New England; being introduced into California, it has become abundant on the Pacific coast.

➤ **Taking It Further: Natural Selection and Complexity**

1. *Answers may vary. The student is asked to determine if the existence and continued growth of simple organisms pose problems for the theory of evolution and explain why or why not.*

2. *Answers may vary. The student is asked to explain why a less complex organism might have a better survival rate than a more complex one.*

Conclusions For Unit 9 (Text page 337)

➤ **Questions: Conclusions For Unit 9**

1. Questions of origin are outside the realm of science because they are not subject to scientific verification, as the testable facts of biology are. Most scientists recognize their limitations as scientists and the limitations of methods of scientific inquiry and do not press for a first cause. Most scholars will agree that first causes extend the searcher beyond the realm of scientific inquiry.

2. *Answers may vary.* Some of the basic assumptions of the scientist and how they are important to scientific study are listed as follows:

   - The scientist begins his work by supposing that there is a real world. Without this assumption, there is no science.
   - The scientist assumes that there is a discoverable *uniformity* and *dependability* about his natural environ-

ment. No hypothesis would be testable without this assumption.
   - The scientist assumes that the unfamiliar is explainable in terms of the familiar through **analogy**. This assumption also is necessary for order in complexity.
   - The scientist assumes that there are simple explanations of things and events. This assumption means that what he discovers is communicable.
   - The scientist assumes that his statements should be subject to criticism and correction. No one person or group of persons can know all there is to know about any given hypothesis, biological function, life process, or species. All must be verifiable by others.
   - The assumption of *causality* is that natural events involve a network of causes and effects. On the basis of knowledge of regular and predictable changes, the scientist detects an association between events and infers a cause-effect relationship.

3. Based on the principles stated above, it is not possible to prove scientifically that either evolution or special creation is the true origin of life. Neither belief is subject to scientific verification.[*]

➤ **Taking It Further: Conclusions For Unit 9**

1. *Answers may vary. The student is asked to explain if the theory of evolution is supported by scientific findings and give support for his conclusions.*

➤ **Questions: Chapter Review**

1. Experienced corn breeders do not attempt to improve a given corn hybrid still further. They only keep up the standard of the inbred strains from which this hybrid was produced. This is done by preventing mixture with other corn and by sorting out any mutations which arise. When better strains are attempted, the breeder goes back and develops them from open-pollinated corn.

2. Clear thinking will show that man's genes have not been changed by use of modern conveniences. There is no means whereby germ plasm is affected by occupation.

3. If corn of the three colors is growing together, select pure white kernels for seed. Plant these on an island or other place where there is no other corn, and you will have a pure white strain. This can be done with white corn because it is recessive; therefore, it has no latent genes for other colors.

Teacher: Question #3 relates directly to Darwin's finches, where recessive genes were allowed to manifest themselves. The only thing that the student may find problematic is that human intervention is required for the above hypothetical scenario.

---

4. Plants having an extra chromosome were found to have some peculiarity but no added vitality. If fact, most polyploids have reduced viability. For exam-

---

[*] The most logical conclusion is: first causes should be studied by all scientists—as well as by all manner of men—through the only reliable source for theological revelation—the *Holy Bible*.

ple, Karpechenko's *cabbage-radish hybrid* had the top of a radish and the root of the cabbage, making it worthless agriculturally; and being sterile, it was unable to reproduce with its parent plants.

5. Darwin did not know the work of Gregor Mendel, the only man of that time who understood genetics. Darwin had been taught that a species does not change in any way. When he saw slight change, which Mendel explained by the coming to expression of latent genes, Darwin went to the opposite extreme from what he had been taught and said that even man has developed by slow changes.

6. It is hard to see how beauty in leaves or stems would give a plant added reproductive ability or protection. Creationists believe that beauty comes from the world having been made with a pleasing variety of beauty. Man can either improve or destroy this beauty.

7. An evolutionist might say, "If you recognize small changes, multiply them by the number of years the earth has existed and you will have large changes." You have learned, however, that there are limits beyond which small changes no longer accumulate. (Recall the breeding of beets, beans, and corn.) There is much breeding of plants and animals which does not involve a change in genes but only in their arrangement. When a gene does change (mutate), there is loss rather than gain.

8. A complex animal or plant does not, because of its complexity, have an advantage in the struggle for existence. Complexity must have been conferred by the Creator rather than by natural conditions such as we observe today.

◆ Think-Session Guide for Unit 9

Theories of Biological Change

Subject: Interpretation of observations

Purpose: To expose the student to teleology

Teacher: In this "search," the student is exposed to the unacceptable "teleological" and the acceptable "functional" interpretations of observations. *Caution*: Do not be lured into "preaching a sermon" before getting into the unit.

a. **To the student:** You may have observed that plants tend to bend toward the light. As a result, more leaf surface is exposed to the light for photosynthesis to occur. With these data, answer the following question: "Why do plants bend toward the light?"

Teacher: Solicit answers until you get one like: "They bend in order to get more light which they need for growth."

Such an answer should lead to a discussion of two points. First, the above answer suggests or asserts intelligent behavior that consciously anticipates future needs. Obviously, there is no reasonably sure data to suggest such a position.

The second point is that, whereas "teleological" interpretations assign conscious purpose to the plant and are therefore unsound, "functional" interpretations are sound and useful. By this we can gain understanding if we associate the action with the result of the action for the organisms. The movement may indeed be interpreted to have survival value or adaptive significance. The plant may be better able to perform some functions having the movement than without having it. Other examples may be given here.

b. **To the student:** Since there are no data to suggest that plants display purpose in their "actions," how would you improve the statement "They bend in order to get more light"?

Teacher: Seek a response something like this: "Plants respond to light by bending toward it; this results in a more efficient use of the available light."

This response is only slightly different from the first. But it should show the difference between a teleological interpretation, which includes outcomes, and the functional interpretation based on data.

Another example may be used:
(Te) Trees make wood to resist the wind.
(Fit) Trees are made of wood, which resists breaking in the wind.

UNIT 10
Ecology and Conservation

CHAPTER 26
Interrelationships of Living Things
(Text pages 341-347)

Understanding God's Creation
Symbiosis
Physical Factors in Environment

◆ **Suggestions for Motivation or Enrichment:**

1. From a well-stocked market make a list of the different kinds of honey which are available from the selective work of bees.

2. Itemize the biological stages involved in producing milk for your breakfast. In discussion with other students, introduce the problems which might interrupt these stages, such as overgrazing of meadowlands, weather emergencies, cattle disease, radioactive fallout, and community discrimination against cattle farming.

3. Write a paragraph to explain what is meant in society by a person being called a "parasite."

4. Keep records of local examples of insects visiting flowers and thus pollinating them. Make your observations by day and by night.

◆ **Suggestions for Multimedia Resources:**[*]

1. *CPC Plant Links.* Center for Plant Conservation. To access this site, visit <http://www.centerforplant-conservation.org/ASP/CPC_PlantLinks.asp#40>. This site provides a lengthy list of links to Web sites about plants, ecology, and conservation.

2. *Centre for Biodiversity and Conservation.* Biology Royal Ontario Museum. To access this site, visit <http://www.rom.on.ca/biodiversity/cbcb/>. Deals with issues related to biodiversity, evolution

◆ **Suggestions for Supplementary Reading:**[*]

1. Horn, C. J. "Water, Water, Everywhere." *Good Science*, October 1997. To read this article online, visit <http://www.icr.org/goodsci/bot-9710.htm>.

◆ **Answers to Questions**

Understanding God's Creation *(Text page 341)*

➤ **Questions: Understanding God's Creation**

1. Ecology is the study of organisms in their relationships to each other and to their environment.

---

[*] The listing of these suggestions does not necessarily imply endorsement of content.

2. It is important to understand ecological principles because man exercises his control over nature best when he understands the intricate balances within the natural world.

Symbiosis *(Text page 345)*

➤ **Questions: Symbiosis**

1. *Examples may vary.* Examples of symbiotic relationships are as follows:

   • Social conjunctive symbiosis occurs when one plant is physically supported by another. Climbers, such as morning glories, and epiphytes, such as orchids, are supported by other plants.
   • Antagonistic nutritive conjunctive symbiosis is a parasitic relationship that occurs when one of the partners is harmed, but the other partner seems to profit from the relationship. The dodder and mistletoe are examples of plant parasites.
   • Reciprocal nutritive conjunctive symbiosis occurs when the relationship is beneficial to both partners. The lichen is an example of this, as it is composed of an alga and a fungus.

2. A food relationship occurs when one or both of the organisms involved get food from the relationship, whereas a social relationship does not involve food.

3. An obligate food relationship is best characterized as conjunctive symbiosis *without permanent contact.*

Teacher: Note that *disjunctive symbiosis* is between organisms that have little to do with each other but live in the same biome—an obligate food relationship is more than this. On the other hand, *conjunctive symbiosis* is between organisms that have close, **permanent** contact, whereas an obligate food relationship does not necessarily involve permanent contact.

---

4. *Examples may vary.* Examples of plant parasites are the dodder and mistletoe. Examples of animal parasites are roundworms and tapeworms.

➤ **Taking It Further: Symbiosis**

1. *Answers will vary. The student is asked to organize the members of a particular community (e.g., a forest, grassland, or desert biome) into a pyramid-shaped food chain.*

Physical Factors in Environment *(Text page 347)*

➤ **Questions: Physical Factors in Environment**

1. The different physical factors and their effects on organisms in their environment are as follows:

   • Gravity: affects the orientation of organisms
   • Light: affects photosynthetic plants in that length of day and intensity requirements are different for dif-

ferent plants; affects animals since they depend on plants for food; affects animals by allowing them to see their surroundings and their prey; also, UV enables some animals to manufacture vitamin D; length of day triggers some instinctive behavior.

- Heat: Most living things exist or are active between 0° C and 100° C. Growth in plants usually stops below and above these temperatures. In animals, hibernation or estivation is often triggered by extreme cold or extreme heat.
- Water: Water is a vital factor in the survival of a given organism in any particular habitat. The amount of water available in a given area will influence which plants and animals live there. For instance, aquatic animals and hydric plants need much water.

2. *Examples may vary.* Some mechanisms in both plants and animals designed to protect them from extremes in these factors are hibernation and estivation in animals, which protect them from cold and heat extremes. In winter, many plants survive as seeds and deciduous plants drop their leaves to prevent water loss. Hydric plants have air channels in the stems, allowing diffusion of oxygen to the roots that are submerged. Some plants, such as cacti, store water for living in dry conditions.

➤ **Taking It Further: Physical Factors in Environment**

1. *Answers will vary. The student is asked to do some research and then describe the effects of all these factors on the African violet.*

➤ **Questions: Chapter Review**

1. Leaf-cutting ants carry pieces of leaf into an underground chamber and place upon them pieces of fungus which grow. This plant is used as food by the ants.

2. Insects which fall into the hollow leaves are digested and absorbed by the pitcher plant. This process gives the plant nitrate, which is abundant in animal tissue.

3. Bees probably are not conscious of their purpose in working. Actually, they are storing food for the bees that will hatch in the spring.

4. Social conjunctive symbiosis is seen in twiners, tendril climbers, leaners, and epiphytes. Examples would be the morning glory, the wild grape, the climbing rose, and the orchid. *Examples may vary.*

5. In a lichen, the alga and fungus are held together by the network of the fungus, hence they are conjunctive. They are reciprocally nutritive in that the fungus holds moisture for the alga, and the latter produces food for the fungus.

6. If left to itself many years, a forest becomes composed of shade-tolerant trees because young seedlings must develop in the shade of mature trees, and intolerant species will not survive.

7. H may unite with OH, giving H_2O, water. *See page 35 of the textbook.*

8. Spanish moss keeps light away by shading the leaves of the tree. (We know the "moss" is not a parasite because many people have seen it growing on wires.)

9. *Answers may vary. The following is a sample experiment:* Fill a glass jar with soil that has only a fair amount of moisture in it and set a small plant in this soil. Every day, place a saturated sponge or piece of wet blanket upon the soil, so that the upper layer of soil has more moisture than the bottom. More roots will grow to the upper layer of soil than to the bottom.

10. Sunny weather is favorable to the production of seed. Moist weather favors production of stems and leaves.

CHAPTER 27
Balance of Nature
(Text pages 349-354)

Soil

Cycles

◆ **Suggestions for Motivation or Enrichment:**

1. Find or carefully expose a cross section of layered soil, heretofore undisturbed. Sketch the layers which are obvious and collect small soil samples for display around the sketch. Postulate the causes of variations.

2. In a plant nursery or garden supply shop note the different kinds of commercial fertilizers. Write down the stated ingredients, with parallel columns of contrast, especially for nitrogen content.

3. Sketch or trace over a local map to emphasize water drainage and apparent relation to various life forms. If the area is flat and arid, mark on your map the artificial water routes.

4. Discuss pH with a swimming-pool serviceman or with laboratory personnel and ask to be shown the instrument for acid-base analysis. Prepare a report on the importance of pH. If possible, include the consideration of pH in the human body.

◆ **Suggestions for Multimedia Resources:**[*]

1. Meng, Alan and Hui. "Food Chains and Webs." *Parenting the Next Generation* (a Christian Parenting Web Site). To read this article online, visit <http://www.vtaide.com/png/foodchains.htm>. Also contains interactive programs, "Create A Food Web" and "Chain Reaction Activity," for students to create their own food web.

---

[*] The listing of these suggestions does not necessarily imply endorsement of content.

◆ Answers to Questions

Soil (Text page 351)

➤ Questions: Soil

1. The different types of soil are as follows:

   - Rock particles form the basis of soil. They are classified according to size as coarse sand, fine sand, silt, and clay.
   - Loam consists of a mixture of various sizes.
   - Humus is the mixture of decayed plant and animal matter found in fertile soil and is essential for plant growth.

2. The beneficial and harmful organisms mentioned in this chapter are as follows:

   - Burrowing animals, such as rodents, can be beneficial because their digging admits air and water to the soil. However, sometimes they admit too much air and the soil dries out.
   - Earthworms are beneficial. Their tunnels admit air and water, and their digestive processes fertilize the soil.
   - Algae are beneficial because they contribute to soil fertility by manufacturing food, form a protective layer at the surface of the soil, and fix nitrogenous compounds in the soil.
   - Fungi help convert humus to nutrients usable to plants and also hold soil particles together, and thus promote water retention.
   - Bacteria are essential in that they are the chief agents that change plant material to humus.
   - Nematodes are harmful to plants.

3. It is important to know the pH of the soil because different plants thrive in different soil pH. It is important to know whether the soil pH is suitable for the plants that will be grown.

4. Mineral composition and pH can be tested using test kits. Observation of the condition of the plants can also give a clue to the condition of the soil.

   Ways to improve the soil are as follows:

   - Earthworms can be added to improve the soil.
   - Mineral fertilizers can add essential minerals such as nitrogen, phosphorus, and potassium.
   - Lime will make a soil less acidic.
   - Peat mold or humus will make alkali soils neutral.
   - Sulfer is added to increase acidity.

➤ Taking It Further: Soil

1. *Answers may vary. The student is asked to list the characteristics of the "perfect" soil for growing plants.*

Teacher: Different plants have different requirements as far as mineral composition, pH, etc. A possible way to approach this question would be to select a particular plant and describe the characteristics of the perfect soil for it.

Cycles (Text page 354)

➤ Questions: Cycles

1. Living things are influenced by cycles. Nature is so balanced that the wastes of one type of organism become the necessary nutrients for another.

2. The basic carbon cycle is as follows:

   The carbon dioxide in the air and soil is taken in by green plants and united with water to form sugar by photosynthesis. An animal eats some part of the plant. Most of the food is united with oxygen in the process of internal respiration. The carbon dioxide is breathed out into the atmosphere, and the cycle is complete.

   Variants of the carbon cycle are as follows:

   The animal may give off some carbon in urine and feces, or the animal may die. The bacteria or fungi of decay then break down the carbon compounds and carbon dioxide is released into the atmosphere.

   Carbon can also be released into the atmosphere through weathering of carbonate rocks and through burning of fossil fuels.

3. A food web is the complete interaction of various food chains in an ecological community. (A food chain is the order of predation of organisms of an ecological community.)

➤ Taking It Further: Cycles

1. *Answers will vary. The student is asked how cycles and food webs illustrate the idea of homeostasis and balance on a wide scale in nature.*

➤ Questions: Chapter Review

1. No. It is only when the other growth factors, such as heat and soil nutrients, are in abundance, and carbon dioxide is the limiting factor, that pumping carbon dioxide into a greenhouse would help.

2. Carbon and oxygen are supplied to plants from the air.

3. Three elements are present in fertilizers more often than others. On the label, the first number is the percentage of nitrogen, the second the percentage of phosphorus, the third the percentage of potassium.

4. Soil with a pH of 8 or any other number above 7 is alkaline.

5. The water cycle, as illustrated in Figure 27-3, goes from one form of water to another and back to the original form; i.e., rain, water in pond; soil water, plant sap, moisture in air, cloud, rain.

6. Carbon from the atmosphere, carbon dioxide, is one part of carbon united with two parts of oxygen.

7. Free nitrogen of the atmosphere is not available to plants except to a few species of bacteria. It is made available by being combined with other elements, forming such compounds as sodium nitrate.

8. By soil organisms we mean plants or animals which live in the soil and change it in some important respect. Examples are bacteria which fix nitrogen from the air, bacteria which decay plant and animal material, molds which do the same, and earthworms which hasten the process.

9. Humus is plant or animal material after it has been decayed. This valuable substance makes the soil easily worked, holds moisture, and supplies the plants with valuable natural fertilizers. (*Every gardener should have a sense of humus! - ed.*)

CHAPTER 28
Biogeography
(Text pages 357-363)

Distribution of Organisms

Biomes

Ecological Succession

◆ Suggestions for Motivation or Enrichment:

1. Make a list of local indigenous plants and prepare a display of names and specimens. Identify their most favored areas on a local map. (Use the USDA Web site <http://plants.usda.gov/index.html> for accurate information; perhaps a plant Web site, such as the *Illinois Natural History Survey* site <http://www.inhs.uiuc.edu/cwe/illinois_plants/>, would be helpful.)

2. Ascertain the local altitude and latitude and identify them in Figure 28-7 of the textbook (page 361). Correlate the local life forms with the chart.

3. Describe a local plant or animal moving by some natural process from the local area until it meets a natural barrier. Discuss the natural barriers, considering the advantages and disadvantages of each one to your community.

4. Research the local needs in the field of soil erosion. Select a topic based on one of the local interest groups, such as housing developers who are involved in hillside grading, small farmers who are victimized by urban encroachment, commuters who want freeways to slash through mountains, and small home owners who cherish normal safeguards for their landscaping.

◆ Suggestions for Multimedia Resources:[*]

1. *Desert Animals & Wildlife.* DesertUSA. To access this Web site, visit <http://www.desertusa.com/animal.html>. Information about desert animals and their habitat; includes pictures

2. Viau, Elizabeth A. *Introduction to Biomes.* World Builders. To access this site, visit <http://curriculum.calstatela.edu/courses/builders/lessons/less/biomes/introbiomes.html>. A review of six different biomes

3. *What's It Like Where You Live?* MBGnet. To access this Missouri Botanical Garden Web site, visit <http://mbgnet.mobot.org/>. An interactive presentation of biomes and ecosystems

4. *Animals A—Z.* Oakland Zoo. To access this Web site, visit <http://www.oaklandzoo.org/atoz/atoz.html>. Describes the animals found in the Oakland Zoo with text and pictures; provides information about animal interdependency and habitat; it has an evolutionary orientation.

5. *Voice of the Deep* (Moody Videos, 30 min.). Life sounds from the "silent" deep—hear a fish croak, a shrimp clack, and a porpoise moo

6. *The Digital Field Trip Series.* Digital Frog International. To access this Web site, visit <http://www.digitalfrog.com/products/index.html>.CD-ROMs; Windows and Macintosh OS. This series offers three virtual field trips to the rain forest, wetlands, and the desert; a workbook is available as part of an educational package. The company has a special home school price for each field trip.

◆ Answers to Questions

Distribution of Organisms (Text page 358)

➤ **Questions: Distribution of Organisms**

1. *Answers will vary.* Examples of several different habitats are as follows:

   • Moist soil is a good habitat for an earthworm.
   • Grassland is a good habitat for sheep.
   • Forests are good habitats for woodpeckers.
   • High mountains are good habitats for mountain sheep.

2. Barriers are important to the distribution of organisms because they prevent the spread of organisms from one area to another. Some examples of barriers are oceans, waterfalls, land masses, and mountains.

3. Some reasons for animal and plant dispersion are that they may disperse to avoid competition with other animals, or physical forces such as winds and streams may carry them, or they may seek to escape their enemies or extremes in weather.

---

[*] The listing of these suggestions does not necessarily imply endorsement of content.

➢ **Taking It Further: Distribution of Organisms**

1. *Answers will vary. The student is asked to discuss the distribution of pine trees after doing some research.*

Biomes *(Text page 362)*

➢ **Questions: Biomes**

1. The more conspicuous plants, such as trees and shrubs, are the indicators of biomes. Temperature zones and variations in rainfall produce the biomes.

2. You would not find reptiles or amphibians in an arctic biome because reptiles and amphibians are cold-blooded. They need environmental warmth to warm their bodies. They would not be active and thrive in an area which was characterized by sub-zero temperatures.

3. Large bodies of water influence the temperatures in a given area. For instance, an ocean can moderate the climate. Large lakes provide an abundant water source, influencing which organisms will thrive in the area. Of course, the ocean itself is a biome (marine biome).

4. Gradients are smooth transitions from one climate or temperature zone to another. The transitions from one biome to another are not usually abrupt. Between the biomes are strips where the plant and animal species from the two biomes intermingle.

➢ **Taking It Further: Biomes**

1. *Answers will vary.*

Ecological Succession *(Text page 363)*

➢ **Questions: Ecological Succession**

1. In a typical succession, some kinds of living things, called pioneers, move into unoccupied territory. Later, these pioneers are crowded out by other organisms. After a time, a reasonably stable condition is reached.

2. Succession is important ecologically because, through succession, habitats are changed. Interactions in nature are almost never static. Through succession, bare areas are replenished. Man, in his interaction with nature, needs to understand the natural progressions of succession, lest he find himself fighting against them.

➢ **Taking It Further: Ecological Succession**

1. *Answers will vary.*

➢ **Questions: Chapter Review**

1. Oceans, mountains, deserts, and other regions where a plant or animal could not live are barriers to its dispersal.

2. Different animals thrive under different conditions. Where one animal can live, another cannot. For example, an ocean is a highway for fish but a barrier for camels. A desert is a barrier for beavers but a highway for camels.

3. A pond or stream surrounded by grass, bushes, worms, and insects would be a good habitat for frogs.

4. Abundant reproduction can result in dispersion as the animals disperse to avoid competition with other animals of like habits.

5. A biome is a large area in which there are plants and animals fitted for the same climate. A habitat is not a certain place but a quality of the environment: the kind of soil and amount of light and heat and moisture which meet the needs of a given plant or animal.

6. The altitudinal gradients of Mt. Kilimanjaro are essentially the same as the latitudinal gradients of the earth from the equator to the polar regions. They both range from tropical rainforest at the equator and at the base of the mountain to tundra in the polar regions and at the top of the mountain. *See page 361 of the textbook.*

7. When the tide is out, little pools of water remain in the low places until the tide comes in again to cover them with water. Tide pools become crowded with bottom dwelling animals such as crabs, mussels, sea anemones, sea urchins, and starfishes. The temporary lack of water causes them to congregate in the pools.

8. Because of the cold temperature, spring comes late on a mountain. For the same reason, autumn comes early.

9. In arid regions, many shrubs have large and extensive roots but small stems and few leaves. The thick roots store water, while the stems and leaves are thin to prevent excess water loss. They may be found where the annual rainfall is twenty inches or less.

CHAPTER 29
Conservation: Applied Ecology
(Text pages 365-373)

Development of Conservation
Solving Problems in Conservation
Planning for Conservation

◆ **Suggestions for Motivation or Enrichment:**

1. Inquire of your town or city government regarding the local sewage-handling procedures and prepare an oral report, including the most urgent problems facing your area.

2. Make a personal tour or investigation of pollution problems in your town or neighborhood and list

them with suggested improvements for presentation to the proper authorities.

3. Write letters to the nearest polluters in your area. Be sure to mention specific observations you have made of pollution from the sources you address.

4. Draw a circle of 50-mile radius around your location on a map and list the wildlife preserves, forest lands, national monuments, and public parks or grounds included in the circle. Select one area and prepare recommendations for the common good.

♦ **Suggestions for Multimedia Resources:**[*]

1. Kasnoff, Craig. *Bagheera In the Wild*. Earth Endangered. (To access the *CKMC* site, visit <http://www.bagheera.com/inthewild>). Highlights endangered or extinct species

♦ **Suggestions for Supplementary Reading:**[*]

1. "Fouling the Nest: Christianity and the Environment." Carl Wieland, *Creation* 24: 110–17 <http://www.answersingenesis.org/creation/v24/i1/fouling.asp>

2. "The Christian and the Greenhouse Effect." Larry Vardiman, Ph.D. *Impact*, June 1990. <http://www.icr.org/pubs/imp/imp-204.htm> copyright 2004, Institute for Creation Research; response of the Christian to the claims of the Greenhouse Effect

3. "Exploring Environmental Ethics." Wayne Wofford, Ph.D. *Scientific Voice*, Center for Scientific Studies, Union University <http://www.uu.edu/centers/science/voice/article.cfm?ID=23>; Christian's environmental responsibility

4. Wenger, Jerome. "The Garden of Eden and Our Environmental Responsibilities as Christians Today." *Papers from the Biology Department Faculty, Covenant College*. To access this article online, visit the Covenant College site at <http://biology.covenant.edu/faculty_work.html>. Dr. Wenger develops the concept of "earthkeeping" along with the Christian's responsibility to God's creation.

♦ **Answers to Questions**

Development of Conservation (Text page 366)
➤ **Questions: Development of Conservation**

1. The warning signs that encouraged conservation were the extinction of the passenger pigeon and the near extinction of the buffalo.

2. The main focus areas of conservation today are the wise use of natural resources and living in harmony with natural balances.

➤ **Taking It Further: Development of Conservation**
1. *Answers will vary.*

Solving Problems in Conservation (Text page 369)
➤ **Questions: Solving Problems in Conservation**

1. The main sources of water pollution are human and industrial wastes, and the main source of air pollution is exhaust pollution from industrial and other sources.

2. *Answers will vary. The student is asked to list some side effects of water and air pollution.* Possible side effects to list would be health problems in humans, health problems or death for animals, and disturbance of the balance in the food web (i.e., one type of organism dies, promoting overgrowth of another).

3. This is not true because man is a member of the biosphere and is a part of its food web. While it is true that some activities of man throw off nature's balance, the tendency of the natural order is to restore balance. From a biblical perspective, man has been placed on earth and given the command to have dominion over the earth, making him not a "problem" to the environment but the steward of it.

➤ **Taking It Further: Solving Problems in Conservation**

1. *Answers will vary. The student is asked to research current exhaust system regulations for cars. Exhaust system regulations vary from state to state. The student should find out the regulations for the area where he lives.* These regulations are for the purpose of reducing air pollution.

Planning For Conservation (Text page 373)
➤ **Questions: Planning For Conservation**

1. It is important to provide a habitat for newly introduced organisms because those organisms will not survive and thrive without a suitable habitat, which includes shelter potential and food supply.

2. Several reasons for conserving plants are that they provide habitats for animals, are potential sources of medicine, are potential sources of genetic material for increased productivity or disease resistance for domesticated plants, and serve to demonstrate what the land was like before the pioneers arrived.

3. Research and technology have helped the conservation effort by developing ways of farming that conserve the soil and water, developing plants with increased productivity and disease resistance, and developing technologies that are less polluting. *Additional answers are possible.*

➤ **Taking It Further: Planning For Conservation**

1. By reducing the large amount of dead wood and undergrowth, the threat of forest fires is greatly

reduced. Also, the opened area would become a habitat for plants and animals that need light. *Additional answers are possible.*

2. *Answers will vary. The student is asked to compare the cost of production of gasoline, ethanol, and blends (E10 and E85) of both fuels.*

➤ **Questions: Chapter Review**

1. In some ponds, especially artificial ones, there are so many fish that none of them grow large. In more recent studies of fish, it has been found that clean water and a good supply of food are more important than restricting the number of fish which are caught.

2. Bacteria consume the oxygen that fish need, so bacteria compete with fish.

3. *Answers will vary. The student is asked to discuss how land has been developed in his area for residential or commercial use with conservation in mind. (For a case in point, visit the following Web site: <prairie-crossing.com>.)*

4. There should be trees and bushes nearby to which the quail can retreat if a hawk or fox appears.

5. *Answers may vary.*

Teacher: The student should have learned by this time that living things affect many people in addition to those people who make money from them.
---

6. *Answers may vary.*
   • **Disadvantages:** possibility of higher taxes; land removed from farming, tax lists
   • **Advantages:** More trees to reduce wind velocity and release oxygen; the best kind of recreation; the satisfaction of conserving God's creation

7. *Answers may vary.* A refuge maintained for the sole purpose of raising game animals may incidentally benefit other animals, especially species that are scarce, imperiled by changes in the environment, or threatened by poachers and predators.

8. Where there are many deer they are detrimental to trees, fruit, and gardens, even pawing out potatoes that have been planted, to get the fertilizer that was planted with them. In forests and parks where their predators have been killed, deer become so numerous that they run out of pasture and starve. Changing the hunting laws and introducing predators into the wild are two of the best ways to manage the deer population.

9. *Answers will vary.*

10. *Answers will vary.* A conservationist would advocate a wise use of resources with attention given to maintaining the natural balance. An environmentalist would oppose any change to the "natural state" of the land.

◆ **Think-Session Guide for Unit 10**

Ecology and Conservation

Subject: Natural selection

Purpose: To present some thoughts on ecology

Teacher: Since so much has been written about this subject, students may think they know it rather thoroughly. However, not all accounts in the press are accurate and reliable. It is best to approach the problem with an open mind.
---

a. **To the student:** Researchers at an experimental fruit farm were testing the effectiveness of EPN (Ethyl p-Nitrophenyl) to control apple mites. When the population reached 10 mites per leaf, half of the block of apple trees was sprayed with EPN. This killed nearly all the mites.

The next week, the population was back up to 8 mites per leaf. Again the test, and half of the block was sprayed, with the same results.

This sequence was repeated for four or five sprayings. At this point, the EPN apparently lost its effectiveness in controlling the mites.

Construct several hypotheses to account for these facts.

Teacher: Elicit suggestions such as: decomposition of EPN; EPN effective only under certain environmental conditions; selective killing of mites most sensitive to EPN. Do not suggest or develop this now.
---

b. **To the student:** In checking their notes, one researcher noticed that only one large batch of EPN was mixed up and that this was used for all sprays. Naturally, he suggested that the EPN was breaking down. Can you suggest two different ways of testing this suggestion (hypothesis)?

Teacher: Note the emphasis on different ways. This will lead to increased reliability of the conclusion drawn. Here, the two may include: 1) Use sprays of different ages on a new block with mites. 2) A chemical analysis of batches of different ages should determine any difference. Other equally valid suggestions may be encouraged.
---

c. **To the student:** A fresh batch of EPN was made up and sprayed on the test half of the block. The population of mites was not significantly decreased.

This fresh EPN was then sprayed on a block that had mites but had not been sprayed. The result was the same as originally observed; most of the mites were killed. Since the two sprays were applied on the same morning, the weather conditions were assumed not to be an influence.

Now let us analyze the problem for its major parts:

1) Material used (the EPN)

2) Conditions at time of use

3) Method of applying

4) Mite population

So far, which of these parts have our hypotheses tested?

Numbers I and 2.

d. **To the student:** The advantage of "listing" the major parts of the problem is that it lets us see all of the possibilities. It often helps to remove any "blinders" we may have around our mind's eye. Which possibilities have we not pursued?

Teacher: Item 3 may be pursued to the teacher's interest. However, item 4 deserves major emphasis.

---

e. **To the student:** Let us take a closer look at item 4, the mites. From your knowledge of the biology in this problem, can you think of what might explain the apparent loss of effectiveness of EPN for controlling the mites?

Teacher: The student may need help here. Recall that after the first spray, most but not all mites were killed. Where did the new population come from? (For our purpose, mites do not migrate.) Were the parents more sensitive or more resistant to the EPN? The trees were sprayed again. If individuals in the population vary in susceptibility, which ones are more likely to survive (and thus reproduce)? Therefore, natural selection, with EPN applying the environmental pressure, resulted in the survival of those mites best adapted to live in the new environment containing EPN.

APPENDIX
A Classification of Organisms

In this appendix, the major groups of organisms are arranged according to a generally recognized system of classification. For each group there is a brief description. In most cases, an estimate of the number of species in the group is given. Common names are used in the examples wherever appropriate.

For the most part, the classification is not carried below the class level and only includes the major phyla. In kingdoms Monera, Protista, and Fungi the classification will predominantly only include the phyla. In several instances, however, examples are given at the family level. Classification is presented also at the order level of some microorganisms, insects, amphibians, reptiles, and mammals. In addition, to illustrate the complexity of classification at the lower levels, a complete classification of the primate order of mammals to the family level is included.

There is disagreement among biologists on the subject of classification. For example, many biologists now use a system of classification that includes a sixth kingdom called the Archaebacteria (substituting Eubacteria for Monera). The system used in this appendix is the standard five-kingdom system.

❦ KINGDOM MONERA

❧ Subkingdom Archaebacteria

✦ PHYLUM HALOPHYLIC BACTERIA
Found in marine (salt water) environments.

✦ PHYLUM METHANOGENIC BACTERIA
Produce methane gas by breaking down organic matter. Often found within the intestinal tracts of animals, in swamps, and in sewage treatment plants.

✦ PHYLUM THERMOACIDIPHILIC
Live in extreme (hot, acidic) environments.

❧ Subkingdom Eubacteria

✦ PHYLUM CYANOBACTERIA (blue-green algae/bacteria)
The simplest algae, mostly aquatic or terrestrial in moist habitats. Single-cells, colonies, filaments or sheets. Sexual reproduction unknown. Blue color due to preponderance of phycocyanin along with chlorophyll a and blue pigment phycocyanin. Stored food similar to glycogen. About 2,500 species. Example: *Nostoc*.

✦ PHYLUM PROCHLOROPHYTA
Use both chlorophylls a and b to carry out photosynthesis. Live in association with certain marine animals. Example: *Prochloron*.

✦ *PHYLUM ANAEROBIC PHOTOTROPHIC BACTERIA*

Without using water as a starting material, these bacteria carry out photosynthesis and do not produce oxygen. Examples: green-sulfur bacteria, purple bacteria.

✦ *PHYLUM SCHIZOPHYTA (heterotrophic eubacteria)*

The bacteria can be classified according to shape: cocci (spherical), bacilli (rodlike), spirilla (spiral). May occur singly or grouped into colonies, or individuals stuck together in chains. Many with limited powers of locomotion by flagella. Complex in molecular composition and structure. About 1,500 species.

❖ *CLASS MICROTATOBIOTES (rickettsia)*

Grow and reproduce only inside living cells.

✶ *ORDER RICKETTSIALES (rickettsia)*

The rickettsiae are parasites of arthropods, but many also produce diseases in man, e.g. Rocky Mountain spotted fever, Epidemic typhus.

❖ *CLASS SCHIZOMYCETES (true bacteria)*

All unicellular, many motile. Can grow and reproduce on a great variety of food sources, both living and nonliving.

✶ *ORDER PSEUDOMONADALES*

Some are photosynthetic or chemosynthetic. Examples: *Thiorhodaceae*, *Vibrio comma* (causes cholera).

✶ *ORDER EUBACTERIALES (true bacteria)*

This order contains the largest number of bacterial species of concern to man. Examples: *Escherichia coli*, *Diplococcus*, *Staphylococcus*, *Streptococcus*, *Bacillus*.

✶ *ORDER ACTINOMYCETALES*

These bacteria have a strong tendency to branch. Occasionally referred to as "mold-like bacteria." Examples: *Mycobacterium tuberculosis* (causes tuberculosis), *Streptomyces*.

✶ *ORDER SPIROCHAETALES*

Cell in form of a spiral, all forms motile. No flagella. Examples: *Treponema pallidum* (causes syphilis).

🦌 KINGDOM PROTISTA

✦ *PHYLUM EUGLENOPHYTA*

The plantlike flagellates. Locomotion by whiplike appendages called flagella. Occur singly or in colonies. Some perform photosynthesis. Many ingest food and are thoroughly animal-like in nutrition; some are parasites. Examples: *Euglena*, *Cryptomonas*.

✦ *PHYLUM DINOFLAGELLATA*

Unicellular algae that have two flagella. The unique feature of this organism is the cellulose and silica plates which provide an armor-like cell wall. Examples: *Ceratium*, *Gymnodinium*, *Noctiluca*.

✦ *PHYLUM ZOOMASTIGINA*

The animal-like zooflagellates. Locomotion by whiplike appendages called flagella. Most unicellular; live inside plants or animals. Some free living in fresh water. Example: *Trypanosoma*.

✦ PHYLUM CILIOPHORA

Animal-like protists; locomotion by means of cilia. The largest and most complexly organized of the protists. Usually with both a large and small nucleus. About 10,000 described species. Examples: *Paramecium, Didinium, Stentor, Euplotes, Vorticella.*

✦ PHYLUM SARCODINA

Animal-like protists; locomotion by means of pseudopods. Many produce intricate shells or skeletal structures; others are naked bits of cytoplasm. Usually divided into five classes. About 8,000 species. Examples: *Ameba, Foraminifera, Radiolaria.*

✦ PHYLUM SPOROZOA

Parasitic protists with complicated life cycles. Usually with no locomotor structures (pseudopods or flagella may be present in certain stages of the life cycles of some). No contractile vacuoles. About 2,000 species. Examples: *Eimeria stiedae* (causes *coccidiosis* in rabbits), *Plasmodium* (carried by mosquitoes, causes malaria in birds, mammals, and man), *Babesia* (carried by tick, causes Texas fever of cattle).

✦ PHYLUM MYXOMYCOTA (acellular slime molds)

A group of less complex organisms with both animal and plant characteristics. Individuals may occur as large, single masses of cytoplasm with hundreds of nuclei and no cell boundaries. This mass moves about and ingests food like a gigantic ameba. Reproduces by spores, shown in some genera to have cellulose walls, formed within sporangia (plant characteristic). Widely distributed, found under damp conditions on decaying vegetation. Example: *Lycogala.*

✦ PHYLUM CHLOROPHYTA (green algae)

The "grass-green algae," mostly aquatic. Single motile or nonmotile cells, motile and nonmotile colonies, filaments, ribbons and tubular forms. Sexual reproduction often by means of motile gametes, but many other methods, both sexual and asexual are known. Cells highly organized. Characteristic pigments chlorophylls a and b, xanthophylls, and beta-carotene, the same as higher plants. Food stored as starch. About 5,700 species. Examples: *Chlamydomonas, Volvox, Closterium, Pediastrum, Spirogyra, Ulva.*

✦ PHYLUM CHRYSOPHYTA (golden algae)

Simple algae in which carotene and xanthophylls are dominant. Photosynthetic products never stored as starch, but often as an oil. Mostly unicellular, but colonial, filamentous, and tubular forms are known. About 5,700 species.

❖ CLASS XANTHOPHYCEAE (yellow-green algae)
Yellow-green in color, otherwise superficially like green algae. Characteristic pigments are chlorophylls a and e, beta-carotene and a xanthophyll. Oil or leucosin usually present in the cells. Examples: *Tribonema, Vaucheria.*

❖ CLASS CHRYSOPHYCEAE (golden-brown algae)
Mostly plankton species, rarely found in large numbers. Characteristic pigments are chlorophyll a, beta-carotene, and two xanthophylls. Motile cells with one longer and one shorter flagellum. Examples: *Synura, Dinobryon, Ochromonas.*

❖ CLASS BACILLARIOPHYCEAE (diatoms)
Cell walls contain silicon compounds and are in two pillbox-like halves. Characteristic pigments are chlorophylls a and c, beta-carotene, and xanthophylls. Examples: *Pinnularia, Navicula, Melosira.*

✦ *PHYLUM PHAEOPHYTA (brown algae)*

Marine algae often with rather complex bodies. Plants usually brownish or olive-green in color. Characteristic pigments are chlorophylls a and c, beta-carotene, and xanthophylls, especially fucoxanthin. Food stored as carbohydrate laminarin. Cell walls often surrounded with algin, an economically important substance. About 900 species known. Examples: *Ectocarpus, Laminaria, Fucus, Macrocystis.*

✦ *PHYLUM RHODOPHYTA (red algae)*

Predominantly marine algae, usually red. Many extremely complex life cycle patterns are known, but there are no motile reproductive cells. Characteristic pigments are chlorophylls a and d, carotenoids, and phycoerythrin. Food stored as a special starch. About 2,500 species known. Examples: *Porphyra, Nemalion, Polysiphonia, Batrachospermum, Lemanea.*

✦ *PHYLUM ACRASIOMYCOTA (cellular slime molds)*

Found in similar locations to the acellular slime molds. Reproduction is asexual. Example: *Dictyostelium.*

✦ *PHYLUM OOMYCOTA (water molds and mildews)*

Consists of highly branched, single-celled filaments. Plant Parasites. Example: *Saprolegnia.*

❦ KINGDOM FUNGI

✦ *PHYLUM ZYGOMYCOTA (the conjugation fungi)*

Mycelia mostly tubular without cross-walls. Terrestrial and aquatic species present. Sexual reproduction by conjugation. Examples: *Synchytrium, Allomyces, Rhizopus,* black bread mold.

✦ *PHYLUM ASCOMYCOTA (the sac fungi)*

Mycelium with definite cross-walls. Mostly terrestrial. Common lichen component. Spores produced in a special saclike cell (the ascus) following a sexual fusion. About 12,500 species. Examples: *Saccharomyces,* yeast, *Penicillium, Neurospora, Aspergillus, Morchella,* morels, *Ceratostomella* (causes Dutch elm disease).

✦ *PHYLUM BASIDIOMYCOTA (the club fungi)*

Mycelium with definite cross-walls; cells often binucleate. Large, terrestrial sporophores common. Spores produced on the surface of a special cell (the basidium) following a sexual process. About 13,000 species. Examples: *Puccinia,* wheat rust; *Ustilago,* corn smut; *Psalliota,* edible mushroom.

✦ *PHYLUM DEUTEROMYCOTA (Fungi imperfecti)*

All fungi in which sexual reproduction is unknown. About 10,500 species. Example: *Trichophytons,* a group of fungi responsible for ringworm of the scalp (*Tinea capitis*), barber's itch, and athlete's foot.

❦ KINGDOM PLANTAE

✦ *PHYLUM BRYOPHYTA (nonvascular plants)*

Small plants without vascular tissue, true roots, stems, or leaves, but often appearing superficially like vascular plants. Sex organs multicellular, with an outer sterile jacket of cells. Gametophyte the dominant generation; sporophytes often parasitic on gametophyte. About 24,000 species.

❖ *CLASS HEPATICAE (liverworts)*

Plants often distinctly liverlike in shape. In many cases with definite dorsoventral symmetry. Sporophytes small, of short duration. About 8,500 species. Examples: *Riccia, Pellia, Marchantia.*

❖ *CLASS ANTHOCEROTAE (hornworts)*

Similar to liverworts in appearance of gametophyte; sporophyte much larger, nearly independent. About 50 species. Example: *Anthoceros.*

❖ *CLASS MUSCI (mosses)*

Plant body not dorsoventrally flattened, plants often leafy. Gametophytes and sporophytes both relatively long-lived. About 15,000 species. Examples: *Sphagnum, Rhodobryum, Mnium.*

✦ *PHYLUM TRACHEOPHYTA (vascular plants)*

Structurally complex, with vascular tissue. Main generation is the sporophyte. Gametophytes often microscopic, parasitic. About 211,900 species.

◆ Subphylum Psilophyta (whisk-ferns)

Leafless, rootless plants represented by just two living genera. Sporangia at ends of branches, but in living forms appearing lateral. Four species. Examples: *Psilotum, Tmesipteris.*

◆ Subphylum Lycophyta (club mosses, quillworts, spike mosses)

Plants with numerous spirally-arranged small leaves. Sporangia in axils of sporophylls, often clustered into cones. Leaf gaps absent. About 1,000 species. Examples: *Lycopodium, Selaginella, Isoëtes.*

◆ Subphylum Sphenophyta (horsetails)

Harsh plants with ridged and grooved jointed stems. Leaves vestigial in small whorls. Leaf gaps absent. Sporangia on highly modified sporophylls; these aggregated into a cone. About 25 species. Example: *Equisetum.*

◆ Subphylum Pterophyta (ferns, gymnosperms, and angiosperms)

Leaves and stems usually larger and more complex; leaf gaps present. About 210,700 species.

❖ *CLASS FILICINEAE (ferns)*

Stems usually underground, leaves complex. Sporangia numerous, mostly clustered lower leaf surfaces. Water required for free-swimming sperm. About 10,000 species. Examples: *Azolla, Marsilea, Botrychium, Asplenium, Dryopteris, Polystichum, Polypodium, Pteridium.*

❖ *CLASS GYMNOSPERMAE (conifers)*

Stems and leaves complex, spores produced in special cones. Seeds not enclosed in a fruit, "naked." Sperm motile or nonmotile, free water not necessary for fertilization, enclosed in pollen tube. Xylem with tracheids only; phloem, no companion cells. About 700 species. Examples: *Cycas, Pinus, Cedrus, Juniperus, Thuja, Ephedra, Welwitschia,* spruce, hemlock, fir.

❖ *CLASS ANGIOSPERMAE (flowering plants)*

Stems and leaves complex, spores produced in a flower. Seeds enclosed in a fruit. Sperms non-motile, enclosed in pollen tube. Number of species greater than all other plant species combined. No resin, 3 types of cells in xylem, phloem with companion cells. The orders of flowering plants are poorly defined. Botanists have concentrated primarily upon the families, and the following are a few common ones out of the 300 described.

✧ Subclass Dicotyledoneae (dicots)

Flowering plants with two cotyledons (seed leaves) in the embryo. Floral organs in groups of fours or fives, leaves netted-veined, stems with a cylinder of vascular tissue, cambium, and pith cells present. About 200,000 species.

✶ Family Ranunculaceae (buttercups)
Herbaceous, usually yellow or white, stamens ten or more, pistils and petals five or more, ovaries superior. Examples: buttercup, marsh marigold, columbine.

✶ Family Cruciferae (mustards)
Petals four, sepals four, ovary superior; stamens in two sets, four long and two short (rarely only two or four); often with a turnip-like or cabbage-like odor. Examples: cabbage, wallflower, radish, turnip.

✶ Family Rosaceae (roses)
Herbaceous plants, shrubs, trees; sepals and many stamens in a ring (hypanthium) surrounding the carpels; leaves usually with stipules; flowers regular, usually perfect; ovaries one to many, superior or inferior. Examples: rose, cherry, plum, apple, strawberry.

✶ Family Leguminosae (legumes)
Flowers usually sweet pea-shaped (a few regular); leaves usually compound; fruit a legume (pea-like pod); herbaceous plants, shrubs or trees; stamens four to ten, commonly diadelphous (nine united by their filaments, one free or nearly so). Examples: bean, pea, alfalfa, clover, locust.

✶ Family Polemoniaceae (phlox)
Herbaceous plants; flowers perfect, regular; corolla of five united petals; stamens five, filaments attached to the corolla tube; ovary superior; styles dividing into three linear stigmas. Examples: phlox, *Polemonium*.

✶ Family Labiatae (mints)
Leaves opposite; stems usually square in cross section; often aromatic; flowers irregular; calyx of five united sepals, commonly two-lipped; ovary superior, usually four-lobed; fruit for little seed-like nutlets; stamens four, in two unlike pairs. Examples: henbit, peppermint, spearmint, horehound, catnip.

✶ Family Umbelliferae (parsleys)
Herbaceous plants; flowers small, generally in a simple or compound umbel; petals five; stamens five; leaves alternate and usually compound; ovary inferior. Examples: parsley, parsnip, carrot, dill.

✶ Family Caprifoliaceae (honeysuckles)
Shrubs or vines with opposite leaves, without stipules; flowers regular or irregular; corolla of united petals; stamens five, inserted on the lobes of the corolla; ovary inferior. Examples: honeysuckle, *Viburnum*.

✶ Family Compositae (sunflowers)
Flowers in clusters appearing like a single flower; petals united; ovary inferior; stamens five, attached to the corolla and united by the anthers. Examples: dandelion, aster, daisy, sunflower.

✧ Subclass Monocotyledoneae (monocots)

Flowering plants with one cotyledon in the embryo. Floral organs in groups of three, leaves parallel-veined, stems with scattered vascular bundles, cambium absent. About 50,000 species.

✶ Family Alismataceae (water plantains)
Herbaceous aquatic or marsh plants; leaves usually broad, petioled, with sheathing bases; three sepals; three petals, separate; carpels several to many; ovaries superior. Examples: *Alisma*, *Sagittaria*.

✶ Family Commelinaceae (spiderworts)
Herbaceous, somewhat succulent plants; leaves alternate; bases sheathing; three persistent and usually green sepals; three petals, usually colored and quickly withering; ovary superior; stamens six. Examples: spiderwort, wandering Jew.

✶ Family Bromeliaceae (pineapples)
Mostly epiphytes with narrow, fleshy leaves. Leaf bases often pitcher-like, trapping large amounts of water. Flowers clustered; regular sepals three; petals three, often brightly colored. Examples: Spanish moss, "air plants," pineapple.

⋆ **Family Liliaceae (lilies)**
Essentially like an amaryllis, except that the ovary is superior. Examples: lily, onion, trout lily, greenbrier.

⋆ **Family Amaryllidaceae (amaryllis)**
Herbaceous plants with three petals and three sepals, similarly colored; stamens six, ovary inferior. Much like a lily except for inferior ovary. Examples: amaryllis, century plant.

⋆ **Family Cyperaceae (sedges)**
Grass-like plants with leaf-sheaths united around the stems; flowers perfect or imperfect, subtended by one or two scales; stamens one to three; ovary superior, sometimes surrounded by a saclike structure. Examples: *Papyrus*, *Carex*.

⋆ **Family Gramineae (grass)**
Herbaceous; leaves usually narrow, composed of the slender blade and stem-enclosing sheath; flowers minute, surrounded by chaffy bracts (glumes, lemmas, paleas); without definite sepals and petals; flowers in spikelets. Examples: bluegrass, bent grass, bamboo, corn, oats, wheat.

⋆ **Family Iridaceae (irises)**
Herbaceous plants; three sepals and three petals, both brightly colored; flowers regular or irregular; stamens three, ovary inferior. Examples: iris, blue-eyed grass.

⋆ **Family Orchidaceae (orchids)**
Perennial herbaceous plants, a few without chlorophyll; ovary inferior; flowers irregular; lower of three petals modified to form a liplike or saclike structure; stamens one or two, united with the style to form a column. Examples: *Cattleya*, *Cypripedium*, *Orchis*, *Habenaria*.

KINGDOM ANIMALIA

✦ PHYLUM PORIFERA (sponges)

Aquatic animals, mostly marine and invariably attached to some solid object; usually colonial or consisting of many individuals fused. Body wall consists of two cell layers. Pores in body wall connected to an internal canal system. Skeleton of spicules or spongin. About 3,000 species divided into three classes.

❖ CLASS CALCISPONGIAE
Sponges with calcium carbonate (lifelike) skeletons, simple canal systems, and small size. Marine. Examples: *Grantia*, *Scypha*, and *Leucosolenia*.

❖ CLASS HYLOSPONGIAE (glass sponges)
Larger and more complex; skeletons of silicon. Marine. Example: Venus's flower basket.

❖ CLASS DEMOSPONGIAE
Sponges with spongin and/or silicious skeleton or no skeleton; complex canal system. Large forms common. Marine and freshwater. Examples: bath sponge, sheep's wool sponge, *Euspongia*, *Spongilla*.

✦ PHYLUM COELENTERATA

Mostly marine animals. Body wall of an outer ectoderm and an inner endoderm, enclosing a saclike digestive cavity with a single opening, the mouth. Radial symmetry. No segmentation. The typical saclike type of body organization occurs throughout but is modified to form either a tubular polyp or a bell-shaped jellyfish. Usually provided with tentacles with highly specialized stinging cells. Over 9,000 species, divided into three classes.

❖ CLASS HYDROZOA
Small single individuals, plant-like colonies, or complex colonial individuals. About 3,000 species. Examples: *Hydra*, *Obelia*, Portuguese man-of-war (*Physalia*), *Velella*.

❖ *CLASS SCYPHOZOA (true jellyfish)*

Polyp stage reduced or absent. Many large individuals. About 200 species. Example: *Aurelia*.

❖ *CLASS ANTHOZOA*

Occur individually or in massive colonies. Polyp form predominates. One group forms cups of calcium carbonate. About 6,000 species. Examples: sea anemone, coral, sea fan.

✦ PHYLUM CTENOPHORA (comb jellies and sea walnuts)

A small group of marine jelly-like animals somewhat more advanced than the coelenterates and sometimes included with them. About 100 species.

✦ PHYLUM PLATYHELMINTHES (flatworms)

Worms much flattened dorsoventrally. Bilateral symmetry. Possess a definite third layer of cells, the mesoderm. Central nervous system developed. Free-living and parasitic species. About 6,000 species divided into three classes.

❖ *CLASS TURBELLARIA*

Free-living flatworms with ribbon-like body, generally marine but some freshwater and terrestrial. Ciliated ectoderm, many mucous cells. About 1,500 species. Example: *Planaria*.

❖ *CLASS TREMATODA*

Parasitic leaf-shaped animals. Usually possess suckers. About 3,000 species. Example: liver fluke.

❖ *CLASS CESTODA*

Parasitic, usually ribbon-like. Possess a scolex with suckers and perhaps hooks. About 1,500 species. Example: tapeworm.

✦ PHYLUM NEMERTEA (ribbon worms)

A small phylum related to the Platyhelminthes. Mostly marine. Small to very large (one species reaches 90 feet). About 500 species.

✦ PHYLUM NEMATODA (roundworms)

Nematodes. Parasitic and free-living. Bilateral symmetry. Digestive system tubular, with mouth and anus. Body cavity present but poorly formed. Body elongate, cylindrical, and usually pointed at both ends. About 10,000 species. Examples: *Ascaris, Trichinella*, elephantiasis pathogen, hookworm.

✦ PHYLUM NEMATOMORPHA (horsehair worms)

A small group of exceedingly long and thin worms living in fresh water and the sea. Parasitic within arthropods during early life. About 200 species.

✦ PHYLUM ACANTHOCEPHALA (spiny-headed worms)

Round worms with a spiny attachment organ, parasitic within the intestines of vertebrates. About 100 species.

✦ PHYLUM TROCHELMINTHES (rotifers)

Microscopic freshwater and marine animals with well-developed body cavity and organ systems. About 2,000 species.

✦ *PHYLUM BRYOZOA (mass animals)*

Small, attached, unsegmented animals, living in both fresh and salt water, though more abundant in the latter. Mouth enclosed in a crown of tentacles; intestine U-shaped, anus near the mouth. About 1,200 known species.

✦ *PHYLUM BRACHIOPODA (lampshells)*

Marine animals with symmetrical bivalve shells within which there is a pair of "arms" bearing tentacles. Superficially resemble the mollusks. About 120 species.

✦ *PHYLUM PHORONIDEA (Sedentary, marine, worm-like animals)*

A small group living within burrows on the sea bottom. About 15 species.

✦ *PHYLUM CHAETOGNATHA (arrow worms)*

A half-dozen genera of small, transparent, marine animals. May occur near the surface in huge numbers. About 30 species.

✦ *PHYLUM MOLLUSCA (clams)*

Soft-bodied, unsegmented animals without jointed appendages. Bilateral symmetry tending to asymmetry (snails). Usually a shell is present (absent or rudimentary in some); gills; mantle and mantle cavity; and a foot. Mostly marine but many are freshwater and some (snails) are terrestrial. About 71,000 modern species have been described, and fossil forms are numerous. Usually divided into five classes.

❖ *CLASS AMPHINEURA*

Mollusks with shell composed of eight overlapping plates (exposed or hidden); no distinct head; foot broad and flat. Marine. About 630 species. Example: chiton.

❖ *CLASS GASTROPODA*

Flatfooted mollusks with distinct head, coiled shells (lacking in some). Aquatic or land. About 55,000 species. Examples: snail and slug.

❖ *CLASS SCAPHOPODA*

Marine. Tusk-shaped shells and foot. A relatively small group. About 200 species.

❖ *CLASS BIVALVA*

Plow-footed mollusks with bivalve shells; no distinct head. Marine and freshwater. About 15,000 species. Examples: clam, mussel, scallop, oyster.

❖ *CLASS CEPHALOPODA*

Marine. With or without shells, well-developed head with eyes; foot modified into tentacles with sucking discs. About 400 species. Examples: nautilus, squid, octopus, cuttlefish.

✦ *PHYLUM ANNELIDA (segmented worms)*

Bilateral symmetry; relatively complex organ systems; non-jointed appendages (if any); body segments more or less similar. Marine, freshwater, or terrestrial. Closely related to the arthropods. About 8,000 species have been described, divided into two minor and three major classes.

❖ *CLASS POLYCHAETA*

Almost entirely marine. Burrowers or tube builders. Usually with paddle-like locomotor appendages (parapodia) on each body segment; head appendages common. Sexes separate. About 3,500 species. Examples: clamworm, parchment worm, sand worm.

❖ *CLASS OLIGOCHAETA*

Many adapted to fresh water and land. Body appendages reduced or lacking. Head not distinct. About 2,500 species. Example: earthworm.

❖ *CLASS ARCHIANNELIDA*

A small group of less complex Annelida.

❖ *CLASS HIRUDINEA*

Freshwater and terrestrial. Predaceous or ectoparasitic. No locomotor appendages; anterior and posterior suckers. About 250 species. Example: leech.

❖ *CLASS GEPHYREA*

Degenerate forms found burrowing along the seashore. Classification doubtful.

✦ *PHYLUM ARTHROPODA*

Complex animals; exoskeleton; jointed appendages; segmentation well developed with a tendency for segments to fuse or to be grouped into definite body regions. Respiration by gills, or trachea, or a modification of these. Ventral nerve chord and open circulatory system. Inhabit all environments. The following classes are usually recognized.

❖ *CLASS ONYCHOPHORA*

Tropical, land-living, wormlike arthropods. Possess a series of paired legs showing weak segmentation; tracheal respiration like insects. Of interest because they combine many annelid and arthropod characteristics. Sometimes listed as a separate phylum. About 80 species. Example: Peripatus.

❖ *CLASS CHILOPODA*

Elongate and superficially wormlike arthropods. Little regional specialization of different segments. One pair of long antennae; one pair of poison-claws. Body flat with one pair of walking legs on each segment of the abdomen. Terrestrial. About 800 species. Example: centipede.

❖ *CLASS DIPLOPODA*

Elongate and superficially wormlike arthropods. Little regional specialization of different segments. One pair of short antennae. Body round with two pairs of walking legs on each segment of the abdomen. Terrestrial. About 6,000 species. Example: millipede.

❖ *CLASS ARACHNIDA*

Arthropods with no antennae and with four pairs of legs; respiration by means of trachea or "book lungs"; segmentation reduced. About 15,000 species. Examples: horseshoe crab, scorpion, spider, tick.

❖ *CLASS CRUSTACEA*

Arthropods with two pairs of antennae; respiration by gills. Most occur in aquatic habitats, but a few occur in moist terrestrial situations. One pair of legs per segment. About 25,000 species. Examples: lobster, crayfish, crab, shrimp, barnacle, pill bug, water flea.

❖ *CLASS INSECTA*

Arthropods with one pair of antennae; three pairs of legs; tracheal respiration. Body usually divided into head, thorax, and abdomen. Many with wings. Primarily terrestrial. The insects. About 700,000 species. Some common orders and their usual characteristics are as follows:

✶ *ORDER THYSANURA*

Small, wingless insects with fishlike settles on body; three long bristles form tail. Example: Silverfish.

✷ ORDER COLLEMBOLA

Minute, wingless insects. Abdomen frequently with ventral jumping organ. Example: spring-tail.

✷ ORDER ORTHOPTERA

Medium to large insects; terrestrial; fore-wings leathery, hind-wings folded fanlike, used for flying; chewing mouthparts. Incomplete metamorphosis. Examples: grasshopper, roach, cricket, mantis.

✷ ORDER ISOPTERA

Small, soft-bodied, antlike insects; wings narrow, equal length; winged and non-winged forms; chewing mouthparts; social. Incomplete metamorphosis. Example: termite.

✷ ORDER ANOPLURA

Small, wingless insects; body flat; sucking or piercing mouthparts. Parasites. Example: lice.

✷ ORDER HOMOPTERA

No wings or two pairs of wings held arched together; wings, same thickness; jointed beak for piercing and sucking attached to base of head. Incomplete metamorphosis. Examples: cicada, leafhopper, aphid, scale insect.

✷ ORDER HEMIPTERA (True bugs)

No wings or two pairs of wings, with forewings partly thickened; jointed beak for sucking arises from front of head. Example: bedbug.

✷ ORDER ODONATA

Fairly large insects with two similar pairs of large, membranous wings, which do not fold; antennae short; body long and slender. Immature forms are aquatic. Chewing mouthparts. Incomplete metamorphosis. Examples: dragonfly, damsel fly.

✷ ORDER EPHEMERIDA

Two pairs of transparent wings; hind wings smaller; long two- or three-pronged tails. Immature forms aquatic. Example: mayfly.

✷ ORDER LEPIDOPTERA

Two pairs of scaly wings; coiled sucking tube for mouthparts. Usually fairly large, showy insects. Complete metamorphosis. Examples: butterfly, moth.

✷ ORDER DIPTERA

Usually two-winged; hind wings may be reduced to knoblike organs of balance; small or medium-sized insects; antennae small; eyes large; sucking and piercing mouthparts. Complete metamorphosis. Examples: fly, mosquito.

✷ ORDER COLEOPTERA

No wings or two pairs of wings, the outer pair a hard sheath beneath which the hind pair is folded; chewing mouthparts; short antennae usually. Largest order. Forewings meet in straight line down back. Metamorphosis complete. Example: beetle.

✷ ORDER HYMENOPTERA

No wings or two pairs of wings hooked together, hind wings smaller; chewing or sucking mouthparts. The only insects with "stingers." Complete metamorphosis. Examples: bee, wasp, ant.

✦ PHYLUM ECHINODERMATA (starfish, sea urchins, sand dollars)

All marine. Complex. Adults radially symmetrical (usually on a plan of five repeated parts); larvae bilaterally symmetrical. Skin usually spiny; body wall with calcareous plates (may be reduced); water vascular system; large coelom. About 5,000 modern species divided into five classes.

❖ CLASS CRINOIDEA

Echinoderms with plantlike appearance. With long arms usually branched and frequently with a stalk attached to ocean bottom. About 635 species. Example: sea lily.

❖ *CLASS ASTEROIDEA*

Echinoderms typically in the form of a five-pointed star with the central disc passing into five regularly spaced projections (arms or rays) with ventral grooves. About 1,500 species. Example: starfish.

❖ *CLASS OPHIUROIDEA*

Echinoderms with five flexible arms (sometimes branched) sharply marked off from the central disc and not ventrally grooved. About 1,500 species. Examples: snake star, brittle star, basket star.

❖ *CLASS ECHINOIDEA*

Spherical or disc-shaped without arms, but with long spines or short, hair-like structures arising from a skeleton of interlocking plates. About 770 species. Examples: sea urchin, sand dollar.

❖ *CLASS HOLOTHUROIDEA*

Elongated echinoderms without arms, but with tentacles around the mouth. Also, without spiny skin, but with microscopic skeletal particles embedded in leathery body wall. About 600 species. Example: sea cucumber.

✦ *PHYLUM CHORDATA (chordates)*

The vertebrates and their invertebrate kin. Phylum is characterized by the possession, at some stage of development, of a notochord; a hollow, dorsal, nerve cord; and serial pharyngeal gill slits or pouches. Segmentation and coelom evident. There are about 70,000 existing species of chordates divided into four subphyla. The first three may be combined under the name Protochordata. The remaining subphylum, the Vertebrata, may be divided into seven classes.

◆ Subphylum Urochordata (tunicates)

Small, marine, free-floating or attached as adults. No endoskeleton; notochord and dorsal part of nervous system degenerate in adult. About 700 species. Example: sea squirt.

◆ Subphylum Hemichordata

Worm-like marine animals forming burrows. Body divided into three distinct regions. Dorsal, hollow nerve cord in collar region. Gill slits in pharyngeal region. No evidence of endoskeleton. Examples: acorn worm, *Balanoglossus* (Genus).

◆ Subphylum Cephalochordata

Small, free-swimming, translucent marine animals, with fishlike body. A well-developed hollow, dorsal nerve cord, notochord, and gill apparatus. About 28 species. Examples: lancelet, amphioxus.

◆ Subphylum Vertebrata

The notochord is replaced by a backbone of vertebrate as the central axis of the endoskeleton. There is an enlarged brain that is protected by a cranium of cartilage or bone. Most have appendages in pairs and nearly all have olfactory organs, ears, and eyes. By far the majority of chordates belong to this subphylum. About 45,000 species. They can be divided into the following classes:

❖ *CLASS AGNATHA (jawless fishes)*

Eel-like fishes lacking paired appendages (fins) and having suction, disc-type mouth. They have a skeleton of cartilage and a two-chambered heart. No scales or jaws. Exposed gill slits. About 10 species. Examples: lamprey, hagfish.

❖ CLASS CHONDRICHTHYES (cartilaginous fishes)

Well-developed cartilaginous skeleton. These are marine fish with true jaws and paired appendages (fins). Two-chambered heart. Uniform vertebrae. Exposed gill slits. About 600 species. Scales. Examples: shark, ray, skate.

❖ CLASS OSTEICHTHYES (pisces-bony fishes)

The majority of modern fish belong to this group that possesses a bony skeleton, scales, and an air bladder. Two-chambered heart. Covered gills. Paired fins. About 20,000 species. Example: mackerel.

❖ CLASS AMPHIBIA

Animals with a moist, smooth, scaleless skin. Larvae typically aquatic, respiring by gills; most adults typically terrestrial and respiring by lungs. Typically a tetrapod (two pairs of appendages), though these may be reduced or lacking in some species; digits without claws. Three-chambered heart. Cold-blooded. About 2,800 species. Examples: salamander, frog, toad.

✶ ORDER APODA (caecilian)

Body slender and wormlike, no feet. Usually in tropics.

✶ ORDER CAUDATA

Two pairs of legs; tails. Larvae usually aquatic. Examples: salamander, newt.

✶ ORDER SALIENTIA

Jumping form, with hind legs larger than forelegs. Tailless. Larvae usually aquatic. Examples: frog, toad.

❖ CLASS REPTILIA

Lung breathing animals, most of which have scaly skins. Independent of water for breeding purposes; eggs with membranes enclosing fluids and with shell. Three-chambered heart (the third sometimes almost completely subdivided). Cold-blooded. About 7,000 species.

✶ ORDER TESTUDINES (or CHELONIA)

Broad body encased in rigid "shell." No teeth, jaws with horny sheaths. Thoracic vertebrae and ribs usually fused to shell. Oviparous, eggs laid in holes or nests. Examples: turtle, tortoise, terrapin.

✶ ORDER RHYNCHOCEPHALIA

Tuatara only living representative; lizard-like. Granular scales, row of low spines down center of back. Lives on land, in water, and in burrows. Lays eggs in holes in the ground.

✶ ORDER SQUAMATA

• **Suborder Sauria (lizards)**
Slender bodies, four limbs, sometimes reduced. Eyelids and tongue movable. Mostly oviparous.

• **Suborder Serpentes (snakes)**
Long bodies, no limbs or feet. No ear openings. Eyes immobile and covered by transparent scales, no lids. Teeth slender and conical, on jaws and roof of mouth.

✶ ORDER CROCODILIA

Long body; large, long head; powerful jaws rimmed with numerous conical teeth. Tongue immovable. Four short limbs ending in clawed, webbed toes. Tail long, heavy, and compressed. Thick, leathery skin with horny plates. Deposits eggs in nests of decaying vegetation. Examples: alligator, caiman, gavial, crocodile.

❖ CLASS AVES (birds)

Feathers. Skeleton with hollow bones and solidified to form rigid, bony box. Warm-blooded or homeothermic (maintains constant body temperature). Four-chambered heart. About 8,600 species.

❖ CLASS MAMMALIA

Possess hair of some kind. Young born alive and fed on milk secreted from mammary glands of mother. Homeothermic (warm-blooded) and Four-chambered heart. About 5,000 species. Some of the common orders of mammals are:

NONPLACENTAL MAMMALS

✴ ORDER MONOTREMATA

Reptile-like mammals found in Australian region. Egg laying; mammary glands without nipples. Examples: duckbill platypus, spiny anteater.

✴ ORDER MARSUPIALIA

Mammals whose young are born immature and then carried in a pouch. Examples: opossum, kangaroo.

PLACENTAL MAMMALS

✴ ORDER INSECTIVORA

Small, insect-eating mammals, adapted for burrowing. Usually with very high metabolism and voracious diet. Examples: shrew, mole, hedgehog.

✴ ORDER DERMOPTERA

Capacity for gliding, with "wings" extending from side of head and including hind limbs. Example: flying lemur.

✴ ORDER CHIROPTERA (bats)

Only flying mammals. Two major groups are the large fruit bats of the tropics and the smaller, insect-eating bats found around the world.

✴ ORDER EDENTATA

Considerable loss or reduction in teeth. This group of fruit and insect-eating New World mammals lives mostly in the tropics. Examples: anteater, armadillo, sloth.

✴ ORDER PHOLIDOTA

Horny plates give appearance of reptile scales. Example: scaly anteater.

✴ ORDER CETACEA

Marine mammals with forelimbs as flippers and hind limbs absent. Skin naked, eyes small, and head very large. Examples: whale, dolphin, porpoise.

✴ ORDER SIRENIA (manatee)

Wholly aquatic herbivores with hind limbs absent and tail broad, flat, and expanded.

✴ ORDER CARNIVORA

Flesh-eaters, usually with sharp claws and highly specialized teeth for catching and tearing flesh. Examples: dog, cat, bear, skunk.

✴ ORDER HYRACOIDEA (hyrax)

Outwardly resemble rabbits. Oddly enough, some believe they are related to the order Sirenia (manatee, Steller's Sea Cow,[*] dugong[**]) and Proboscidea (elephants), as shown by dentition, toenails, sensitive pads on their feet, small tusks, and the shape of some of their bones.

✴ ORDER PERISSODACTYLA

Plant-eating, hoofed, and grazing mammals. With an odd number of toes; one, three, or five. Examples: horse, zebra, tapir.

✴ ORDER ARTIODACTYLA

Hoofed plant eaters with an even number of toes. Usually with horns or antlers. Examples: pig, sheep, deer, cattle, hippopotamus.

---

[*] Steller's Sea Cow (*Hydrodamalis gigas*) was a large sirenian mammal formerly found near the Asiatic coast of the Bering Sea. It was discovered in the Bering Strait in 1741 by the naturalist Georg Steller and became extinct in 1768.

[**] Exodus 35:7 (NIV) states, "ram skins dyed red and hides of sea cows[a]; acacia wood; ..." The footnote [a] states that the term for *sea cows* refers to *dugongs*; this term is also found in verse 23.

✷ *ORDER RODENTIA*

A large worldwide group of mostly small mammals with teeth adapted for gnawing; four large incisor teeth with sharp cutting edge. High reproductive capacity. Examples: rat, mouse, squirrel, muskrat, beaver.

✷ *ORDER TUBULIDENTATA (aardvark)*

Tubular teeth. Eats ants.

✷ *ORDER LAGOMORPHA*

Mammals similar to rodents but with an extra pair of incisors, peglike teeth. Examples: rabbit, hare.

✷ *ORDER PROBOSCIDEA (elephant)*

Herbivorous mammals with specialized noses (trunks) and tusks.

✷ *ORDER PRIMATES*

Much enlarged cranium, with the eyes rotated to front of head. They usually stand erect, have thumbs opposing the fingers and usually with nails replacing claws. Examples: monkey, ape, lemur, tarsier.

The complexity of classification is shown in the following example, a complete classification of the order Primates to the family level.

- **Suborder Prosimii**
  Infraorder Lemuriformes
  Superfamily Tupaioidea

  ✷ **Family Tupaiidae (tree shrews)**
  Superfamily Lemuroidea

  ✷ **Family Lemuridae (lemurs)**

  ✷ **Family Indridae (lemurs, indris)**
  Superfamily Daubentonioidea

  ✷ **Family Daubentoniidae (aye-ayes)**
  Infraorder Lorisiformes

  ✷ **Family Lorisidae (lorises, pottos, galagos)**
  Infraorder Tarsiiformes

  ✷ **Family Tarsiidae (tarsiers)**

- **Suborder Anthropoidea**
  Superfamily Ceboidea

  ✷ **Family Cebidae (New World monkeys)**

  ✷ **Family Callithricidae (marmosets)**
  Superfamily Cercopithecoidea

  ✷ **Family Cercopithecidae (Old World monkeys, baboons)**
  Superfamily Hominoidea

  ✷ **Family Pongidae (apes)**

Ad maiorem

Dei gloriam

Christian Liberty Press

502 West Euclid Avenue

Arlington Heights, Illinois 60004

www.christianlibertypress.com

ISBN-13:   978-1-930367-93-7
ISBN-10:   1-930367-93-7

9 781930 367937